Julian Kellen

Dof

Advisory Committee on
Microbiological Safety of

Interim Report on Campylobacter

Advises the Government on the Microbiological Safety of Food

London:HMSO

First published 1993

ISBN 0 11 321662 9

CONTENTS

CHAPTER 5
SOURCES AND TRANSMISSION OF INFECTION

CHAPTER 6
CAMPYLOBACTER IN ANIMALS

CHAPTER 7
CAMPYLOBACTER INFECTIONS IN HUMANS: POSSIBILITIES FOR PREVENTION

THE COMMITTEE AND ITS TERMS OF REFERENCE

The Parliamentary Under Secretary of State for Health announced in the House of Commons on 12 June 1990 the establishment of the Advisory Committee on the Microbiological Safety of Food. The Committee was given the following terms of reference:-

> "To assess the risk to humans of micro-organisms which are used or occur in or on food and to advise Ministers on the exercise of their powers in the Food Safety Act 1990 relating to the microbiological safety of food".

ADVISORY COMMITTEE ON THE MICROBIOLOGICAL SAFETY OF FOOD

LIST OF MEMBERS

CHAIRMAN

Professor Heather M Dick
Professor of Medical Microbiology, University of Dundee

MEMBERS

Mr R Ackerman	Chairman of Hotel and Catering Training Company
Mr G Amery	Formerly Group General Manager, Technical Group, CWS Ltd.
Dr C St J Buxton	Director of Public Health, Durham Health Authority
Professor R Feldman	Professor of Clinical Epidemiology, London Hospital Medical College
Professor D Georgala	Director, Institute of Food Research
Dr R Gilbert	Director of Food Hygiene Laboratory and Deputy Director of Central Public Health Laboratory, Public Health Laboratory Service
Dr P Mullen	Veterinary Adviser to Union International Ltd.
Dr S Palmer	Regional Consultant Epidemiologist, Communicable Disease Surveillance Centre, Welsh Unit, Cardiff Royal Infirmary
Ms B Saunders	Freelance consultant to consumer groups
Dr N Simmons	Consultant Bacteriologist and Head of Department of Clinical Bacteriology, Guy's Hospital, London
Mr R Southgate	Technical Director, Northern Foods Meat Group
Mr R Sprenger	Director of Environmental Services, Doncaster Metropolitan Borough Council
Dr G Spriegel	Director of Scientific Services, J Sainsbury PLC
Dr M Stringer	Director of Food Science Division, Campden Food and Drink Research Association
Dame Rachel Waterhouse	Formerly Chairman of Consumers Association
Dr T Wilson	Senior Consultant Bacteriologist, Belfast City Hospital

ASSESSORS

Mr R Alexander	Welsh Office
Mr B Bridges	Department of Health
Dr R Cawthorne	MAFF
Mr E Davison	Scottish Office Agriculture and Fisheries Department
Dr H Denner	MAFF
Mr B Dickinson	MAFF
Dr G Jones	Department of Health
Dr J Ludlow	Welsh Office
Dr A MacLeod	Scottish Office Home and Health Department
Dr C McMurray	Department of Agriculture, Northern Ireland
Dr E Mitchell	Department of Health and Social Services, Northern Ireland
Dr N Peel	Public Health Laboratory Service

SECRETARIAT

Mrs S Gordon Brown	Department of Health
Dr D Harper	Department of Health
Dr D Lees	MAFF
Dr R Mitchell	MAFF
Dr C Swinson	Department of Health

ACKNOWLEDGEMENTS

The Committee wishes to record its thanks to the Public Health Laboratory Service Communicable Disease Surveillance Centre, the Communicable Diseases (Scotland) Unit and the Department of Health and Social Services (Northern Ireland) for supplying information and many of the figures and tables reproduced in the report; and to the many Heads of Departments of British educational establishments, the AFRC Institute of Food Research, the independent Research Associations, the Central Veterinary Laboratory, and the Public Health Laboratory Service Headquarters Co-ordination Unit for providing information about their research activities on *Campylobacter* species. The Committee also wishes to record its appreciation to Dr V King of the Department of Health for her contributions to the report throughout its development.

SUMMARY

1. In this report we have attempted to establish what is known about *Campylobacter*, its natural habitat, sources, routes of transmission and occurrence in animals and in food. We have also tried to identify how it causes human infection, and the extent of the public health problem this poses. In addition we have attempted to identify the gaps in knowledge which need to be filled and the appropriate action required to reduce the problem.

2. We began by looking at the microbiological properties of *Campylobacter* which are most relevant to human disease. Although *Campylobacter jejuni (C.jejuni)* and *Campylobacter coli (C.coli)* are the most common species, current reporting does not usually specify either species or sub-types, as speciation and sub-typing methods are not routinely used in clinical laboratories. Also some non-culturable forms are not detectable by normal laboratory techniques. We concluded that in order to isolate all the species which cause human infection, laboratories need to use culture methods which closely mimic the environmental conditions found in their natural habitat, which is the intestinal tract of a range of wild and domestic warm blooded animals and birds. More information is also needed on the species and sub-types of the organism which cause human infection to enable sources and routes of transmission to be traced.

3. We went on to look at the disease caused by human *Campylobacter* infection (campylobacteriosis). *Campylobacter* has become the most commonly reported cause of acute gastrointestinal infection in the UK, with 43,886 cases reported in 1992. As few as 500 bacterial cells can initiate intestinal colonisation and result in disease symptoms. In the UK symptoms vary from a mild attack of diarrhoea lasting 24 hours, to a severe illness lasting more than a week. Characteristic of campylobacteriosis is persistent colicky abdominal pain, and the diarrhoea may often be blood-stained. Although uncommon, clinical complications can occur. Some *Campylobacter* strains, particularly those found in the UK, appear to cause damage by attachment and then invasion of the cells lining the gut wall. Other strains, isolated in developing countries, may cause damage as a result of the production of toxins. The existence and extent of immunity or resistance to *Campylobacter* infection in the UK is unknown. By the detection of a specific immune response, it may be possible to determine the fraction of the population that has been exposed to *Campylobacter*. Higher levels of serum antibodies in people who are regularly exposed to *C.jejuni* seem to indicate a degree of immunity.

4. We looked at current surveillance, which is based on voluntary reporting by laboratories to the Public Health Laboratory Service Communicable Disease Surveillance Centre (PHLS CDSC) in England and Wales, with similar systems in Scotland and Northern Ireland. Data trends have shown a steady increase in the number of cases reported and *Campylobacter* is now the most commonly reported cause of gastrointestinal disease in all parts of the UK. *Campylobacter* thus poses a significant public health problem. The true extent of the problem is

unknown, as it is thought that not all cases present to their general practitioner, or have stool samples examined. Most reported cases are apparently sporadic, outbreaks are uncommon. The reasons for this are unknown. Peak incidence occurs in young adults. Reported cases peak in late spring, often with a secondary peak in the autumn. There are also regional variations in the numbers of cases reported. The reasons for these age, seasonal and regional variations are not fully understood.

5. It is known that *Campylobacter* is commonly found on poultry meat, in raw milk, in sewage, in untreated water, and can sometimes be found on red meat carcases and offal. Contact with pets with diarrhoea, particularly puppies, has been identified as an additional risk factor, as has milk from bottles pecked by birds. The relative importance of these different sources in causing human infection is unknown. Cross contamination of foods can occur by direct contact, or indirectly via hands or kitchen equipment such as chopping boards. *Campylobacter* is, however, killed by chlorination, pasteurisation and proper cooking. Reports of person to person transmission are uncommon.

6. Intestinal colonisation by the species of *Campylobacter* which usually cause disease in humans is common in a wide variety of animals and birds, but in contrast to humans, rarely results in disease. In poultry it has been shown that eggs from breeding flocks are free from *Campylobacter* infection, with chicks becoming colonised when they are 3-5 weeks old. It has been suggested that contamination from the environment accounts for this, either by transmission on the clothing or footwear of farm personnel, or from rearing houses which have not been effectively cleaned and disinfected between flocks, or from contaminated water consumed by the chicks. Poultry slaughtering processes can also provide opportunities for cross contamination, via scalding tanks, plucking machines and automated evisceration equipment, or through immersion chilling of carcases. Air chilling of carcases could reduce contamination but may be inefficient because of the surface and texture of poultry skin. Red meat slaughtering processes, by contrast, may reduce opportunities for *Campylobacter* contamination of carcases.

7. Finally we looked at ways of controlling the risk to humans of *Campylobacter* infection. At the present time campylobacteriosis is generally considered to be a foodborne infection for which foods of animal origin constitute the most important source. Control and prevention therefore ultimately depend on reducing the numbers of *Campylobacter* and other microorganisms in the whole food chain, although this is a long term goal.

8. In the shorter term a more practical approach is to consider the growth and survival characteristics of *Campylobacter* in foods, to form the basis for recommending practical measures that can be taken now by industry and consumers to prevent *Campylobacter* infection. Such an approach is used by industry in the Hazard Analysis and Critical Control Point (HACCP) system to identify those points in the food chain where numbers of organisms can be

controlled or reduced. We examined available information on the effects of storage temperature, heating, salt levels, water activity, acidity/alkalinity, atmosphere and other factors. We concluded that storage at below 30°C, salt levels in some foods, drying, and a pH of less than 4.9 would all help to inhibit growth of the organism, but that only heat treatment sufficient to kill vegetative cells will destroy campylobacters.

9. Cross contamination of foods is thought to be an important means of transmitting *Campylobacter* to humans. The importance of handling raw food with care, to avoid direct or indirect cross contamination to ready to eat foods, cannot be stressed enough.

10. We believe that in order to be effective the above measures must be combined with general food hygiene training and education for food handlers and the public.

11. At the end of each Chapter we have made recommendations, a number of which address topics for further research. Our conclusions and recommendations are drawn together at the end of the report.

CHAPTER 1

INTRODUCTION

Introduction

1.1 This report from the Advisory Committee on the Microbiological Safety of Food (ACMSF) reviews *Campylobacter* infection in humans and animals and sets out interim advice on the measures which may be taken to reduce the incidence of campylobacteriosis. The ACMSF intends to produce a further report on *Campylobacter* when the results of surveillance and research work recommended in this report are available.

General background to *Campylobacter*

1.2 The role of campylobacters as human pathogens was not fully appreciated until 1977 when routine laboratory methods for isolation were developed and reporting began. Since then the reported numbers of laboratory diagnosed cases have steadily increased in all parts of the United Kingdom (UK), and in a number of other European countries *Campylobacter* has also been recognised as a significant cause of infectious intestinal disease.[1] However, the differences in the systems of statutory notification, epidemiological investigation and reporting, as well as different levels of under-reporting, mean that it is difficult to make direct comparisons between countries.

1.3 In the UK reported numbers of laboratory diagnosed cases have exceeded those of *Salmonella* since 1981 in England and Wales and since 1985 in Scotland. In 1991 reports of *Campylobacter* isolates in Northern Ireland also exceeded those of *Salmonella*. Sources of infection of *Campylobacter* and *Salmonella* have much in common. Incubation period, disease symptoms and their duration are similar, but infectious dose is currently considered much smaller for *Campylobacter* than *Salmonella*, and there are significantly more deaths due to *Salmonella* infection. Furthermore, there are seasonal differences between infection with the two pathogens, with the major peak of *Campylobacter* infection occurring in late spring, 6-8 weeks before that of *Salmonella*. The temperatures and atmospheres under which campylobacters survive and grow also differ significantly from those of salmonellae. A more detailed comparison of the biological properties of *Campylobacter* and *Salmonella* species associated with human enteritis is set out in Table 1.1.

1.4 *Campylobacter* is now the pathogen most frequently isolated from cases of acute gastrointestinal infection throughout the UK. In 1992, the last year for which figures are available, the numbers of reported cases were 38,552 in England and Wales, 4,915 in Scotland, and 419 in Northern Ireland.

Table 1.1
COMPARISON OF THE BIOLOGICAL PROPERTIES OF *CAMPYLOBACTER* AND *SALMONELLA* SPECIES ASSOCIATED WITH HUMAN ENTERITIS

	CAMPYLOBACTER	*SALMONELLA* (non-typhoid)
Incidence (number of cases (lab. reports)/ 100,000 population)*		
Eng/Wales 1992 (1991)	75.7 (64.1)	62.1 (44.5)
Scotland 1992 (1991)	97.0 (67.7)	58.7 (45.7)
N. Ireland 1992 (1991)	26.4 (19.2)	14.1 (10.0)
Mortality (number of deaths)**		
Eng/Wales 1992 (1991)	2+ (2)	N/A (62)
Scotland 1992 (1991)	0+ (0)	0+ (2)
N. Ireland 1992 (1991)	0+ (0)	0+ (0)
Seasonal Peak(s)	Late Spring (May) and, most years, Autumn (September)	Summer (August)
Disease symptoms	Bloody diarrhoea, rarely vomiting, fever may be present	Diarrhoea and vomiting, fever, headaches
Normal duration of symptoms	1-7 days	1-3 days
Incubation period ***	2-5 days	6-72 hours
Infectious dose (bacterial cells)	< 500	Usually $> 10^5$
Sources of infection	Raw meat, in particular poultry, raw milk, contaminated water and milk, pets with diarrhoea	Raw meat, in particular poultry, eggs, raw milk
Growth at ambient temperature (22°C)	No	Yes
Temp. range for growth	30-45°C	5-45°C
Minimum pH for growth	5.0	4.0
Minimum A_w for growth	0.98	0.94
Growth in Atmospheric O_2	No	Yes
Sensitive to heat (pasteurisation, normal meat cooking)	Yes (non-sporing)	Yes (non-sporing)

* See footnotes to Table 4.1
** CDSC (England and Wales), CD(S)U (Scotland), DHSS(NI) (Northern Ireland)
*** Time between ingestion of an infectious dose and appearance of symptoms
\+ Provisional N/A Not Available

1.5 There is no doubt therefore that *Campylobacter* poses a significant public health problem in the UK. The economic cost of this is considerable. A study in 1986 estimated that each identified case in England cost society and the individual £587, a total of £14 million in that year.[2] Furthermore, not all cases present to their doctor or are microbiologically confirmed if they do, so the size of the problem and total cost is probably much greater.

1.6 Although the report of the Committee on the Microbiological Safety of Food (the Richmond Committee) highlighted the role of *Campylobacter* in causing foodborne disease, it did not consider campylobacters in any detail.[3, 4] In the light of this, and the significance of *Campylobacter* as a cause of acute infectious diarrhoea, the ACMSF decided as one of its early priorities to review the information available on the role of campylobacters in causing foodborne disease in humans, and report to Ministers.

The ACMSF's approach to its work

1.7 The Committee first sought to establish the size of the problem within the UK by considering the data available on the incidence of *Campylobacter* infection. We then considered wide-ranging information on possible sources and routes of transmission and research funded by the Health and Agriculture Departments and by British academic establishments.

1.8 We then considered the scope for improving the safety of food in the food chain (e.g. transport, slaughter, processing, handling and packaging) and what advice can currently be given to food handlers and the public to help reduce the number of cases of foodborne *Campylobacter* disease.

1.9 We finally considered the need for further research.

CHAPTER 2

THE INFECTIVE AGENTS

Introduction

2.1 The bacteria that we now refer to as *Campylobacter* were first isolated in 1913 by McFadyean and Stockman,[5] and were originally thought to be pathogens of veterinary importance (see Chapter 6). The first identification of *Campylobacter* in the stools of humans with acute enteritis was in 1972.[6] Previously the organism had only been observed in blood cultures.[7] This chapter highlights the microbiological properties of the *Campylobacter* species that are of most relevance to human disease.

Growth and survival

2.2 Campylobacters are slender, spirally curved bacterial rods. Their name is derived from the Greek *campylo* meaning curved and the Greek *bacter* meaning rod. Under certain environmental conditions the bacterial cell becomes round and this change in shape may be associated with a transition from a viable, culturable bacterial form, to a viable, non-culturable form which is not detected by conventional laboratory techniques.[8] This type of transition may aid survival of campylobacters in water,[9] and poultry can be infected by the organism in this form.[10, 11]

2.3 Although some of the newly described campylobacters are able to grow in air, most members of the genus *Campylobacter* require less oxygen than is present in the normal atmosphere ie they are microaerobic. This is because oxygen at normal atmospheric pressure is toxic to their growth.[12] The optimal atmospheric composition for growth of *C. jejuni* has been found to be 5% oxygen, 10% carbon dioxide and 85% nitrogen.[13] As a consequence, bacterial numbers will start to decline on exposure to the atmosphere. This has important implications when considering conditions needed for survival of campylobacters in sources of infection, and during transmission of infection from the source to the susceptible individual.

2.4 Most *Campylobacter* species associated with human enteritis grow well at 42^{0}C and 37^{0}C but not at 25^{0}C and are referred to as the thermophilic group. This means that they will not grow at normal room temperature or refrigeration temperatures. The requirement for a temperature greater than 30^{0}C is probably a consequence of adaptation to conditions found in their normal habitat, which is the intestinal tract of a wide variety of wild and domestic warm-blooded animals and birds[14] (see also Chapter 6).

2.5 Although thermophilic, campylobacters are also heat-sensitive and are inactivated at 48^{0}C. Therefore they would not be expected to survive typical meat cooking procedures or pasteurisation of milk. Campylobacters do not survive drying or acidic conditions, salt concentrations in excess of 2% are inhibitory, and the organisms are sensitive to the bactericidal effects of chlorine.[15, 16, 17] Detailed information from a survival model for *C.jejuni* is now available from the MAFF predictive modelling database (see Appendix 1). These survival characteristics will affect the types of food implicated in transmission of campylobacteriosis and are considered further in Chapter 7.

Isolation and identification of *Campylobacter* species associated with human enteritis

2.6 Since 1977 laboratories have been using the microaerobic and thermophilic properties of *C.jejuni/coli*, in addition to culture medium containing antibiotics, as selective characteristics for their isolation from clinical samples, usually stool specimens.[18] Such samples may contain large numbers of campylobacters ($>10^6$/g) but also many other bacterial species.[19] The growth of other bacteria is inhibited by using selective media and appropriate environmental conditions which encourage the growth of campylobacters. Foods may contain fewer campylobacters so that an additional enrichment procedure is required for their isolation.[20, 21]

2.7 Although it is possible to differentiate *Campylobacter* species by studying a range of biochemical and growth characteristics (see Appendix 1), currently clinical laboratories do not routinely perform this differentiation. Consequently, most laboratories only report "*Campylobacter* species" to the Public Health Laboratory Service Communicable Disease Surveillance Centre (PHLS CDSC) (see Chapter 4). In addition, the antibiotics incorporated into media used for the selective isolation of most *Campylobacter* species, including *C.jejuni*, have the desired effect of inhibiting other enteric bacteria, but also inhibit the growth of some less common *Campylobacter* species such as *C.coli* and *C.upsaliensis*. Thus these species may be under-represented in the epidemiological data.[22]

2.8 In the period 1989-1990, 17% of campylobacters reported to CDSC by laboratories in England and Wales were reported by species, for the remainder only the genus name was given. Where the species was stated 89.5% were *C.jejuni*, 10.3% were *C.coli* and 0.2% were other named species (Report to the ACMSF from CDSC August 1991). The more recently described but less commonly isolated species include *C.lari, C.hyointestinalis, and C.upsaliensis*.[23, 24, 25, 26, 27]

2.9 For epidemiological tracing of sources of infection and transmission routes, sub-typing is necessary to identify the particular strain of *Campylobacter* involved (see Appendix 2). The most commonly used methods for sub-typing *C.jejuni/coli* to date have been biotyping and serotyping.[28, 29, 30] Bacteriophage typing is less commonly used.[31] Reference typing facilities in the UK are restricted to the examination of strains from outbreaks or defined epidemiological studies. Currently there is no national *Campylobacter* reference laboratory; serotyping is performed at Manchester PHL and biotyping and 'phage typing at Preston PHL. If a national reference laboratory became available it could act as a centre for further investigation of *Campylobacter* strains, such as sub-typing. In addition, such a centre could document and store strains for future use.

Use of sub-typing schemes

2.10 Some studies have investigated the origin of *Campylobacter* infection by comparing the biotypes and serotypes of strains isolated from human cases of enteritis with those isolated from animal and environmental sources. One such study found that serotypes commonly found in human infections were frequent among strains isolated from environmental and animal sources.[32] In the UK between January 1982 and December 1988 a comparison was made of the most common Penner serotypes isolated from humans and chickens. The results showed that 71% had the same serotypes[33] (see also Chapter 5).

2.11 Another study assessed the advantages of using three different typing schemes, Penner and Lior serotyping and biotyping, on strains from three outbreaks. One outbreak was caused by the consumption of contaminated cows' milk, another by the consumption of contaminated goats' milk and the third outbreak was associated with puppies. In the cows' milk outbreak it was possible to identify the outbreak strain from humans and from milk samples using both serotyping schemes but not by biotyping alone. In the second outbreak involving the consumption of goats' milk, biotyping was useful because serotyping by either method did not give enough information. In the third outbreak associated with puppies, a combination of any two of the three schemes was able to pinpoint the putative source strain.[34]

2.12 It is clear that current methods used for sub-typing *Campylobacter* are limited in their usefulness because of their inability to discriminate between some strains. In addition, a percentage of strains are untypable by Lior serotyping, Penner serotyping and 'phage typing.[35]

2.13 Newer methods for sub-typing *Campylobacter* are currently being evaluated; these involve the analysis of bacterial DNA and are referred to as genotyping methods (see Appendix 2 for details). An advantage of genotyping methods, unlike serotyping and 'phage typing, is that it is always possible to obtain a result. Genotyping methods also generally provide good discrimination between strains. Some of these methods, like ribotyping, appear to have the advantage of being easier to interpret while retaining the power to discriminate.[36]

2.14 It is not clear at the moment which method or combination of methods would best provide an epidemiologically useful sub-typing scheme and be suitable for use by routine laboratories.

2.15 In order to evaluate all the available sub-typing methods, a comprehensive comparative study needs to be undertaken. Until identification of strains for epidemiological purposes is improved, conclusive evidence of sources and transmission routes will be lacking.

Conclusions

2.16 The significant microbiological properties of most *Campylobacter* species are that they do not grow at normal room temperatures or at refrigeration temperatures, neither do they grow in air. Campylobacters are sensitive to drying, and to acidic conditions. Salt levels over 2% inhibit growth, and the organisms are inactivated at temperatures of 48°C and over. Campylobacters have been shown to exist in a viable but non-culturable form, in which they are not isolated on standard culture media. It is suggested that such forms are a response to certain environmental conditions, but it is not clear whether they are infectious for humans. **(C2.1)**

2.17 There is a need to be able to differentiate between *Campylobacter* species and sub-types to enable better identification of sources of infection. Differentiation of *Campylobacter* species is not currently done by laboratories routinely, so the full significance of different species or sub-types in causing human disease is unknown. **(C2.2)**

2.18 Various sub-typing methods have been used, and new methods are being developed, but some strains are either untypable or indistinguishable using current methods. Other strains can only be accurately identified using a combination of available methods. As there is currently no central *Campylobacter* Reference Laboratory, this work lacks a focal point. **(C2.3)**

Recommendations

2.19 A central UK *Campylobacter* Reference Laboratory should be established to co-ordinate further investigation of *Campylobacter* strains, and we recommend that the Government consider how best this might be done. **(R2.1)**

2.20 In addition, we recommend that Government funds research to expand current knowledge of *Campylobacter* species, in particular:

- to establish isolation and identification methods that can be used by clinical laboratories for the detection of all clinically relevant *Campylobacter* species;

- to develop methods of sub-typing which will enable better epidemiological tracing of sources and transmission routes of human infection; and,

- to develop better detection methods for viable, non-culturable forms of *Campylobacter* in order to determine whether they play a role in the production of human enteritis. **(R2.2)**

CHAPTER 3

DISEASE DESCRIPTION AND IMMUNE RESPONSE IN HUMANS

Introduction

3.1 As set out in Chapter 2, the role of *Campylobacter* species as human pathogens was not fully appreciated until 1977 when selective media were developed for routine laboratory isolations, and reporting began. Since then the number of reports has increased steadily, and *Campylobacter* species are now the most commonly reported cause of infectious intestinal disease exceeding those of *Salmonella* in all parts of the UK. This chapter describes the disease known as *Campylobacter* enteritis or campylobacteriosis, the mechanisms by which campylobacters cause disease and aspects of the host's immune response to these organisms.

Infectious dose and clinical effects

3.2 Human volunteer experiments have shown that ingestion of a dose as small as 500 bacterial cells can produce disease symptoms. Neither the incubation period nor the severity of disease was affected by the size of the dose, but the attack rate among the volunteers increased with an increase in the number of bacteria given.[37, 38, 39] This implies that when an equally small dose is ingested by a group of individuals, not all of them will become ill. Whether an individual becomes ill or not may be determined by a number of complex interacting physiological factors including the immune status of the individual to *Campylobacter*. If most infection occurs as a result of ingestion of a small dose as might be expected because of growth and survival characteristics (see Chapter 2), this could account for the apparently sporadic rather than outbreak nature of most *Campylobacter* infection described below (see Chapter 4).

3.3 Clinical features of *C.jejuni/coli* infection vary from a mild attack of diarrhoea lasting 24 hours to a severe illness lasting more than a week. The most common clinical manifestation of *C.jejuni/coli* infection in developed countries is an inflammatory form of enteritis. The onset of diarrhoea, which is often blood-stained, is preceded by malaise and a fever of 40°C may develop. Characteristic of campylobacteriosis is a persistent colicky abdominal pain which may mimic acute appendicitis. Other symptoms that may be present are headache, backache, aching of the limbs and nausea, but rarely vomiting.[18, 40, 41]

3.4 At the moment there is insufficient data to determine if *Campylobacter* infection without diarrhoea occurs in the UK. From studies reported so far in the UK on healthy controls, asymptomatic faecal excretion appears to be uncommon, <1.0%.[42] In a recent report from Norway convalescent carriage of *Campylobacter* for more than one month was detected in 16% of patients.[43]

3.5 In developing countries asymptomatic faecal excretion is more common, occurring in up to 17% of Bangladeshi children age 1-5 years. The clinical picture in these areas, especially in young children less than 2 years of age, is characterised by profuse watery diarrhoea. Symptomatic infection in adults in developing countries is rare.[42, 44]

3.6 Complications of campylobacteriosis are uncommon but in some individuals illness may be prolonged and severe. In these cases antibiotic treatment may be necessary.[45,46]

3.7 Although complications are uncommon, bacteraemia (bacteria in the blood stream) has occasionally been reported in intestinal infections, but is probably more frequent than has been described.[47] In approximately 1% of cases reactive arthritis has occurred and, more rarely, neurological complications such as Guillain-Barré syndrome (GBS).[41,48] It has been suggested that GBS is more likely to occur after infection with some strains of *C.jejuni/coli* than others, in particular Penner serotype 19.[49,50] Generally, extra-intestinal infections such as urinary tract infection, or inflammation of the gall bladder, pancreas or meninges are rare, but they have been documented in the elderly, immunocompromised or very young.[51,52,53]

3.8 *C.jejuni* infection during pregnancy tends to be mild and self-limiting. *Campylobacter* infection during the first 6 months of pregnancy has resulted in spontaneous abortion, stillbirth, prematurity and neonatal sepsis but these are rare occurrences.[54,55] Advice on avoiding foodborne infections in pregnancy is given in the Department of Health's booklet "While you are pregnant: Safe eating and how to avoid infection from food and animals".[56]

Disease causing mechanisms

3.9 Most studies designed to investigate how campylobacters cause disease have used *C.jejuni* and *C.coli*, which are the commonly isolated *Campylobacter* species associated with human enteritis. The incidence and disease causing potential of the more recently described species, *C.lari, C.upsaliensis,* and *C. hyointestinalis*, has yet to be fully established.

3.10 In order to cause infection, after ingestion, campylobacters must pass the acid barrier of the stomach and gain access to the mucosal surfaces of the lower gastrointestinal tract.

3.11 The disease processes induced by campylobacters appear to be a by-product of growth of the bacteria or colonisation of the human gut. Conversely, campylobacters have been found to colonise the gastrointestinal tract of a wide variety of wild and domestic animals without causing detectable illness or pathology (see Chapter 6). Thus it would appear on the available evidence that campylobacters have greater disease causing potential in humans compared to animals.

3.12 Colonisation of the lower gastrointestinal tract appears to be a pre-requisite for human infection. The ability of campylobacters to move (motility) may help the bacteria overcome the protective barrier of the overlying intestinal mucus gel and gain access to the gut epithelial cells which line the intestine.[15,57] Indeed, the ability of campylobacters to move towards the gut cells may be considered a virulence factor, as both animal and human studies indicate that motile strains colonise the intestine more successfully than non-motile strains.[58,59,60,61]

3.13 The inflammatory form of *Campylobacter* enteritis, which is characterised by blood-stained stools, suggests that the bacteria have invaded the gut epithelial cells causing damage.[62] It has been suggested that some *Campylobacter* strains are more likely to invade and damage the human gut, which might mean that some strains have a greater disease causing potential in humans, but this needs to be substantiated.

3.14 Finally, there is evidence for the role of a cholera-like toxin produced by *Campylobacter* strains in causing the secretory form of enteritis seen in developing countries.[63, 64] In addition to the cholera-like toxin, other *Campylobacter* toxins with different modes of action have been reported.[65, 66] Several other investigators have been unable to find toxin production by *Campylobacter* strains.[44, 67, 68] As with other enteric pathogens, toxin production may be complex and the exact number of distinct toxins produced by *Campylobacter* has yet to be determined. However, toxin producing strains do not usually appear to be associated with the inflammatory diarrhoea most common in the UK.

Aspects of the immune response to *Campylobacter*

3.15 The existence and extent of immunity or resistance to *Campylobacter* infection in the UK population is unknown as no comprehensive surveys have been done. The importance and need for such information in providing a complete picture of the epidemiology of *Campylobacter* infection is clearly demonstrated in the analysis of the National Case Control study results (see Chapter 5).

3.16 Information on the immune response to *C.jejuni* is available from studies done in other countries. These show a rising level of specific serum antibodies of the IgG, IgA and IgM classes detected in patients with *Campylobacter* enteritis.[44, 69] The detection of *C.jejuni* specific IgA and IgM appears to indicate a recent infection, while the presence of IgG alone is indicative of long term immunity from a previous infection.

3.17 People who experience multiple exposures to *C.jejuni* as a result of their occupation, such as slaughterhouse workers or butchers, and those who habitually drink raw milk, have higher levels of serum antibodies than controls and lowered attack rates when exposed to re-infection.[70, 71] In a human volunteer study when re-infection was attempted, the presence of *C.jejuni* specific serum antibodies was associated with a state of resistance to re-infection.[39]

3.18 The higher incidence of intestinal carriage of *Campylobacter* in people in developing countries may indicate that exposure to *Campylobacter* at an early age induces protective mechanisms in the gut, which prevent disease.[72] This protection has been associated with a particular type of antibody, intestinal secretory IgA, which can be detected in gut contents.[73] Laboratory experiments support the theory that secretory IgA prevents campylobacters attaching to intestinal epithelial cells.[74] The presence of *C.jejuni* specific secretory IgA in human milk correlates with the prevention of *Campylobacter* enteritis in breast-fed babies.[75] Further details about the immune response can be found in Appendix 4.

Conclusions

3.19 Campylobacter infection can be caused by ingestion of a small number of bacterial cells. **(C3.1)**

3.20 Some individuals may have a degree of acquired immunity to *Campylobacter* infection, but the extent of such immunity in the UK population is unknown. **(C3.2)**

3.21 The disease symptoms in the UK are of an inflammatory type of campylobacteriosis, and differ from the secretory type of enteritis reported in developing countries, which may be due to infection by toxin producing strains. **(C3.3)**

3.22 The relative potential for different strains of *C.jejuni/coli* and the less commonly isolated species to cause disease in humans is not well understood at this time, and more information is needed. **(C3.4)**

Recommendations

3.23 We recommend that Government funds research to provide more information:

- to establish whether all strains of *C.jejuni/coli* from whatever source have equal disease causing potential for humans;

- to establish the disease causing potential of the more recently described species, *C.lari, C.hyointestinalis* and *C.upsaliensis*;

- to investigate the role of toxin-producing strains in the UK; and,

- to establish the level of immunity to *Campylobacter* in the UK. **(R3.1)**

CHAPTER 4

EPIDEMIOLOGICAL SURVEILLANCE IN HUMANS

Introduction

4.1 This chapter considers the epidemiological surveillance of *Campylobacter* infection in humans in the UK.

Surveillance

4.2 In England and Wales surveillance of *Campylobacter* infections is based on voluntary weekly reports of isolates made to the PHLS CDSC in London by PHLS laboratories, around 300 NHS laboratories and a number of private laboratories. CDSC guidance to laboratories instructs them to report only the first positive isolation from an individual, so the number of reports should represent the number of individuals suffering from campylobacteriosis. Information relating to these laboratory confirmed cases is entered into a computer database at CDSC. Since 1 January 1989 laboratory reports have included the name, age, sex and brief clinical details of cases.

4.3 Outbreaks of *Campylobacter* infection are reported to CDSC by medical microbiologists, consultants in communicable disease control (CCDCs) or are identified by CDSC from laboratory reports. Since 1 January 1992 whenever a general outbreak (i.e. an outbreak affecting members of more than one private residence, or residents of an institution) of campylobacteriosis or other infectious intestinal disease is identified, CDSC sends a blank summary report form to the appropriate CCDC to be filled in when the investigation is complete. This form seeks information on the numbers of individuals at risk, the numbers of cases bacteriologically confirmed, the suspect vehicle (if any) and the nature of the evidence supporting the suspicion. CDSC also ask that any full reports of the investigation be forwarded to them.

4.4 Analogous voluntary laboratory and outbreak reporting systems exist in Scotland and Northern Ireland. In Scotland NHS/university laboratories make weekly reports of isolates and Consultants in Public Health Medicine (Communicable Diseases) report outbreaks to the Communicable Diseases (Scotland) Unit (CD(S)U). In Northern Ireland the 16 hospital laboratories make reports of isolates and CCDCs report outbreaks to the Department of Health and Social Services (DHSS(NI)).

4.5 The Royal College of General Practitioners runs a sentinel (ie representative location) practice scheme for a wide range of diagnoses including infectious intestinal disease, but the reports do not specifically record *Campylobacter*. Statutory notification of food poisoning does not require the organism to be specified, nor is *Campylobacter* separately notifiable. Neither of these sources therefore currently provide data on which to verify the laboratory reporting systems.

Trends in data

4.6 In England and Wales the number of reported laboratory diagnosed cases of campylobacteriosis increased steadily from 12,168 in 1981 to 34,552 in 1990, and the number of cases per 100,000 population rose from 24.5 to 68.1 in the same period. Reports in 1991 returned to the 1989 level, but in 1992 rose again in line with the trend established in the 1980s to 38,552, equivalent to 75.7 per 100,000 population (see Table 4.1, Figure 4.1). Since 1981 *Campylobacter* has been more prevalent than *Salmonella* in England and Wales and is the most frequently isolated pathogen from cases of acute infectious diarrhoea (see Figure 4.2).

4.7 A similar picture has been seen in Scotland (see Table 4.1), although reports of *Campylobacter* isolates did not exceed those of *Salmonella* until 1985. The number of laboratory diagnosed cases rose from 1,887 in 1981 to 4,915 in 1992, equivalent to a rise in rate per 100,000 population from 36.4 to 97.0.

4.8 In Northern Ireland, although there has been a steady increase in the number of reported isolates since 1981 (see Table 4.1), the annual total in 1992 of 419 gives a rate of incidence of 26.4 per 100,000, only about 30% that of the rest of the UK.

4.9 *Campylobacter* therefore poses a significant public health problem. Furthermore, not all cases present to their doctor, or are microbiologically confirmed if they do, so the total extent of the problem is probably much greater. A small study of a defined general practice population suggested that the mean annual incidence of *Campylobacter* infection was about 1.1%.[76] If this were representative of the population it would suggest that over half a million cases occur each year in England and Wales alone. However, these data are difficult to interpret because of varying trends of infection from year to year and in different parts of the country (see paragraph 4.13 below). They emphasise the need for larger population based studies of gastrointestinal infections if the true incidence is to be ascertained. We note that following a pilot study, in July 1993 the Steering Group on the Microbiological Safety of Food began a study in England to establish the incidence and aetiology of infectious intestinal disease in the community and in those attending GP surgeries. We look forward with interest to seeing the results of this study.

Table 4.1
***CAMPYLOBACTER* SPECIES**

LABORATORY REPORTS OF FAECAL ISOLATES TO CDSC, CDS(U) AND DHSS(NI) 1981-1992

YEAR	ENGLAND AND WALES NO.	ENGLAND AND WALES RATE/100,000 POPULATION*	SCOTLAND NO.	SCOTLAND RATE/100,000 POPULATION*	NORTHERN IRELAND NO.	NORTHERN IRELAND RATE/100,000 POPULATION*
1981	12168	24.5	1887	36.4	19	1.2
1982	12797	25.8	1922	37.2	24	1.6
1983	17278	34.8	1895	36.8	45	2.9
1984	21018	42.2	2181	42.4	58	3.7
1985	23572	47.2	2563	49.9	90	5.8
1986	24809	49.5	2372	46.3	73	4.7
1987	27310	54.4	2740	53.6	122	7.7
1988	28761	57.1	2906	57.0	173	11.0
1989	32526	64.3	3080	60.5	192	12.1
1990	34552	68.1	3625	71.1	244	15.4
1991	32636	64.1	3430	67.7	306	19.2
1992	38552	75.7	4915	97.0	419	26.4

* Based on the estimated mid-year population for each year extrapolated from the 1981 census for the years 1981-90, and the projected mid-year population from a 1989 base for the year 1991. Projected population for 1991 used for 1992.

Sources: Laboratory reports - CDSC (England and Wales), CD(S)U (Scotland), DHSS(NI) (Northern Ireland)
Population figures - Office of Population Censuses and Surveys (England and Wales), General Register Office (Scotland) and General Register Office (Northern Ireland), as quoted in Central Statistical Office Annual Abstract of Statistics 1992

Figure 4.1

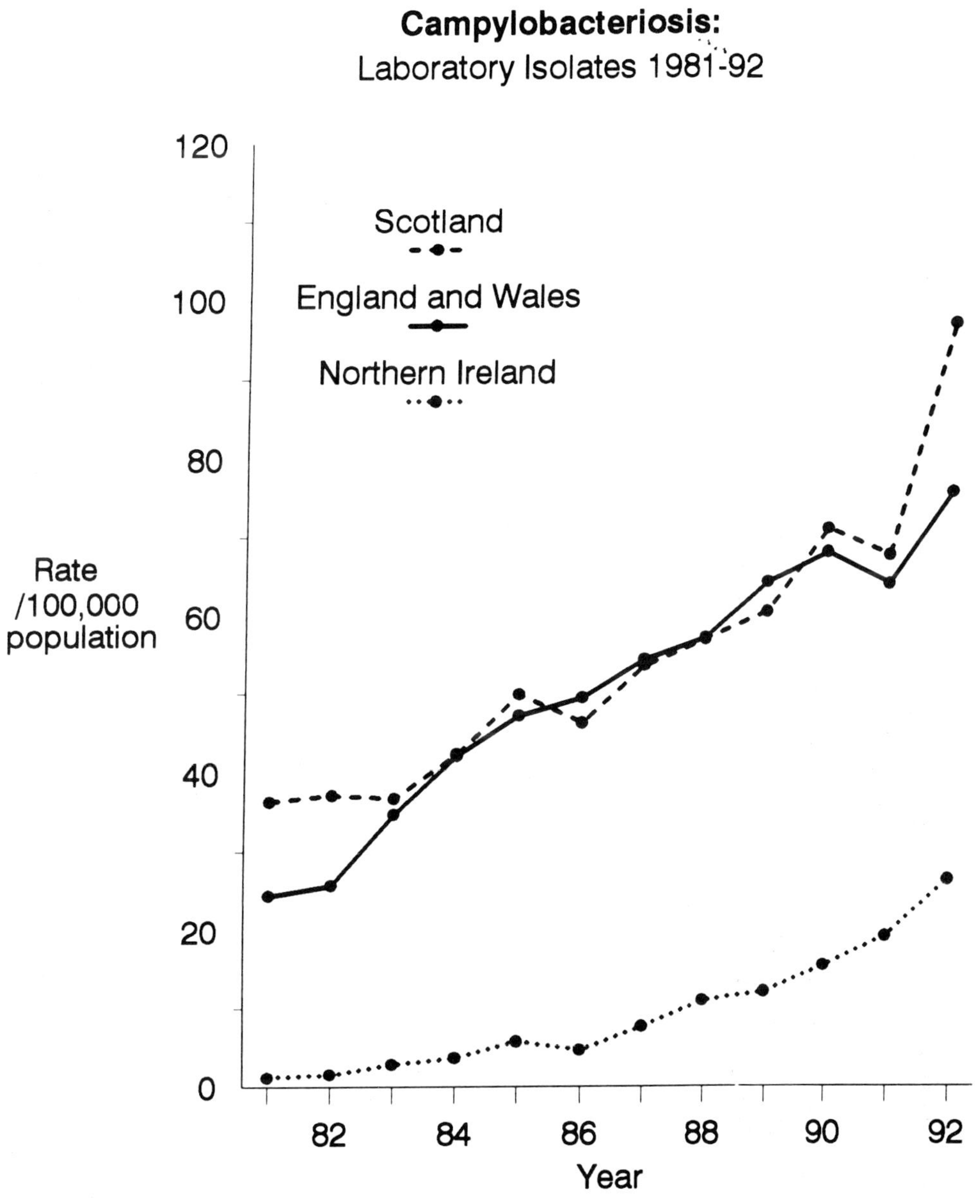

Source: PHLS-CDSC, CD(S)U, DHSS-NI
Prepared by Department of Health.

Figure 4.2

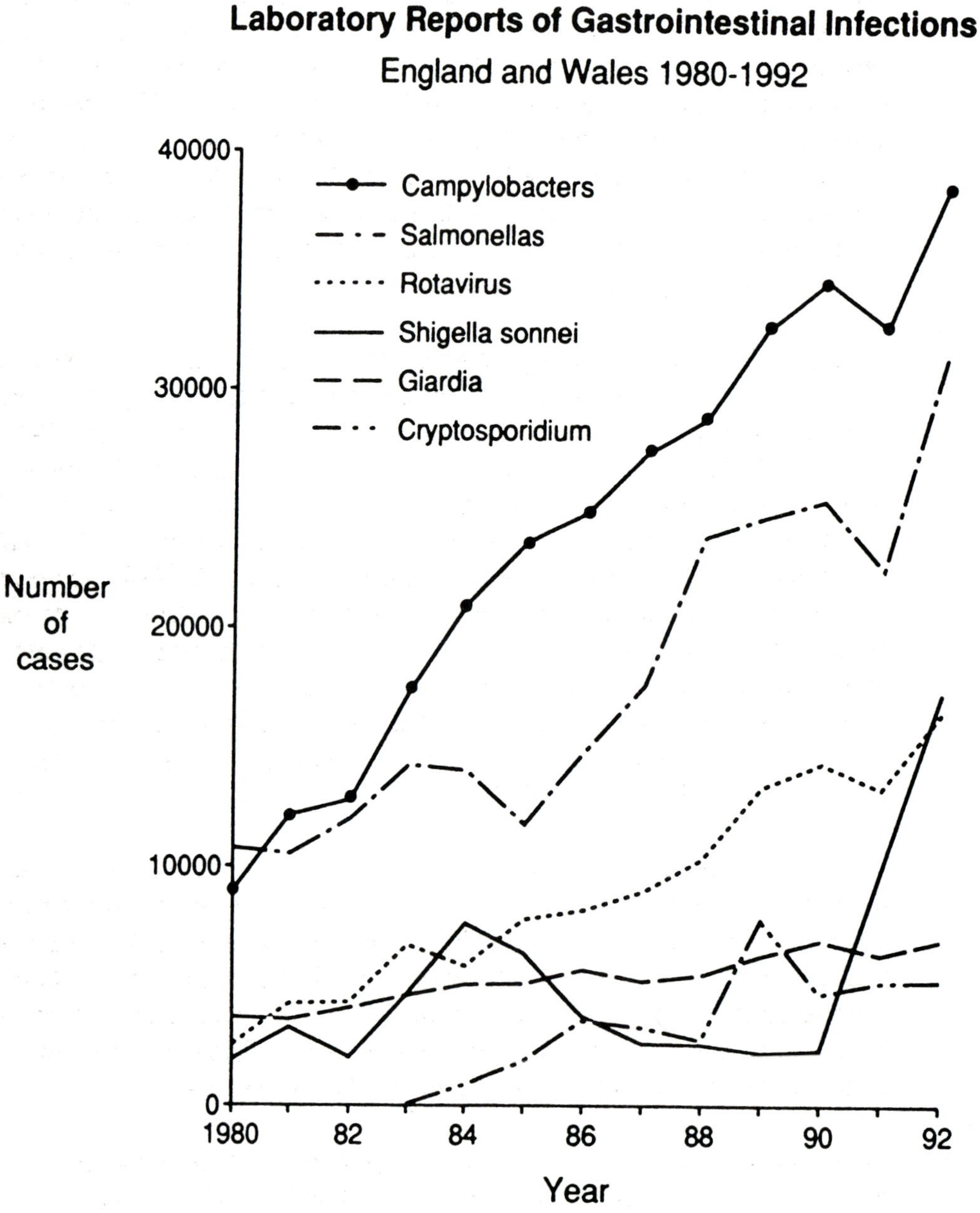

Source: PHLS
Prepared by CDSC

Descriptive epidemiology of *Campylobacter* infections

4.10 The vast majority of cases of *Campylobacter* infection are apparently sporadic, and outbreaks are rarely identified. For example, in England and Wales only about 10 general outbreaks of *Campylobacter* infection were identified per year between 1988 and 1992 (see Table 4.2). Furthermore, although the improved data handling and storage system introduced in 1989 resulted in a marked increase in the ascertainment of household outbreaks (i.e. outbreaks affecting members of the same private residence only, and not connected with any other case or outbreak), fewer than 1% of cases reported to CDSC in 1992 were part of known outbreaks, whether household or general. The reasons for the apparently sporadic nature of *Campylobacter* infection are unknown. One possible explanation has already been given (see Chapter 3). On the other hand, apparently sporadic cases may in fact represent unrecognised outbreaks which might be more fully identified if current sub-typing systems were more widely applied and sub-typing was developed further.

4.11 In England and Wales *Campylobacter* appears predominantly to be a disease of young children and young adults, in particular males (see Figure 4.3). This is similar to the pattern which has been noted in Scotland (data from CD(S)U) and Northern Ireland (data from DHSS(NI)). However as these findings are based on reports of faecal isolates from patients with diarrhoea they will be affected by any differences in the sampling rates from different age groups. A PHLS survey undertaken in five laboratories in England in 1983 and 1984 showed sampling rates were disproportionately high in 1-4 year olds.[77] Using faecal samples as denominator, the lowest incidence of any age group was in infants, and the highest incidence was in young adults, particularly young men. Similar findings have been reported following a further two year study starting in 1985 and involving 16 laboratories in England and Wales (PHLS Study Group 1990. Report to ACMSF from CDSC).

4.12 Analysis of laboratory reports in England and Wales has shown that the seasonal incidence of *Campylobacter* varies with a marked peak in late spring, and in most years a secondary peak in the autumn (see Figure 4.4). The reasons for this are unknown. A similar pattern is seen in Scotland and Northern Ireland. This seasonality contrasts with that for *Salmonella* which peaks some 6-8 weeks later. This suggests that either the sources are different for the two pathogens or that different factors are important for their transmission.

4.13 There are also regional variations in the reporting of *Campylobacter* infections (see Figure 4.5) which require further investigation. These may reflect a higher incidence in rural than urban populations as has been reported in other countries, but may also be associated with varying modes of transmission.

Table 4.2

***CAMPYLOBACTER* INFECTION**

OUTBREAKS REPORTED TO CDSC
ENGLAND AND WALES
1988-1992

OUTBREAKS	1988	1989	1990	1991	1992+
Total	15	360	328	398	446
General	10	9	11	8	10
Family	5	351	317	390	436

\+ Provisional figures

Note: An improved data handling and storage system was introduced in 1989. This improved ascertainment of household outbreaks from 1989 onwards.

Source: PHLS CDSC

Figure 4.3

Campylobacter enteritis
Incidence rates
England and Wales, 1992

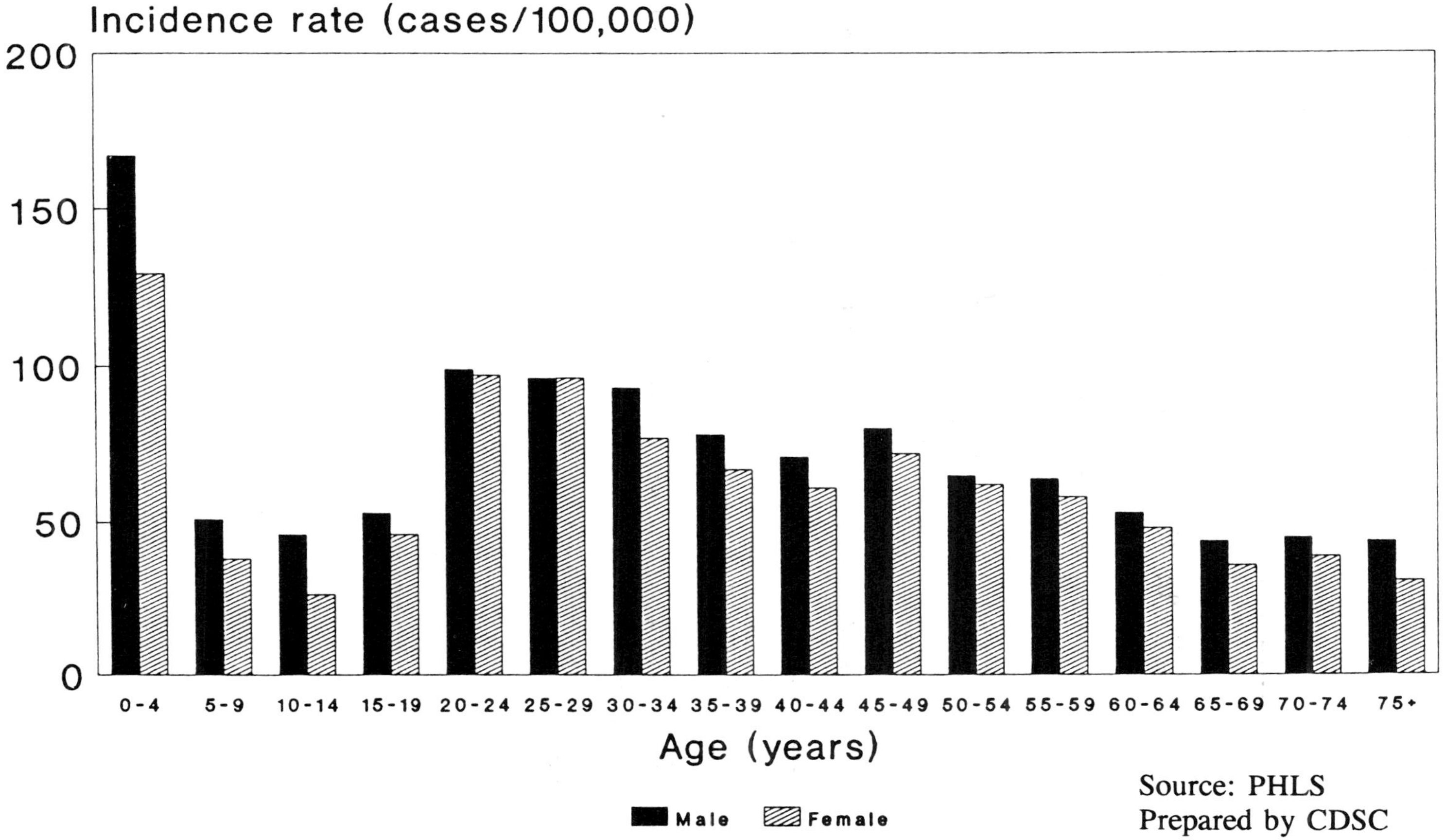

Source: PHLS
Prepared by CDSC

Figure 4.4

Campylobacter enteritis

England and Wales 1978-1992

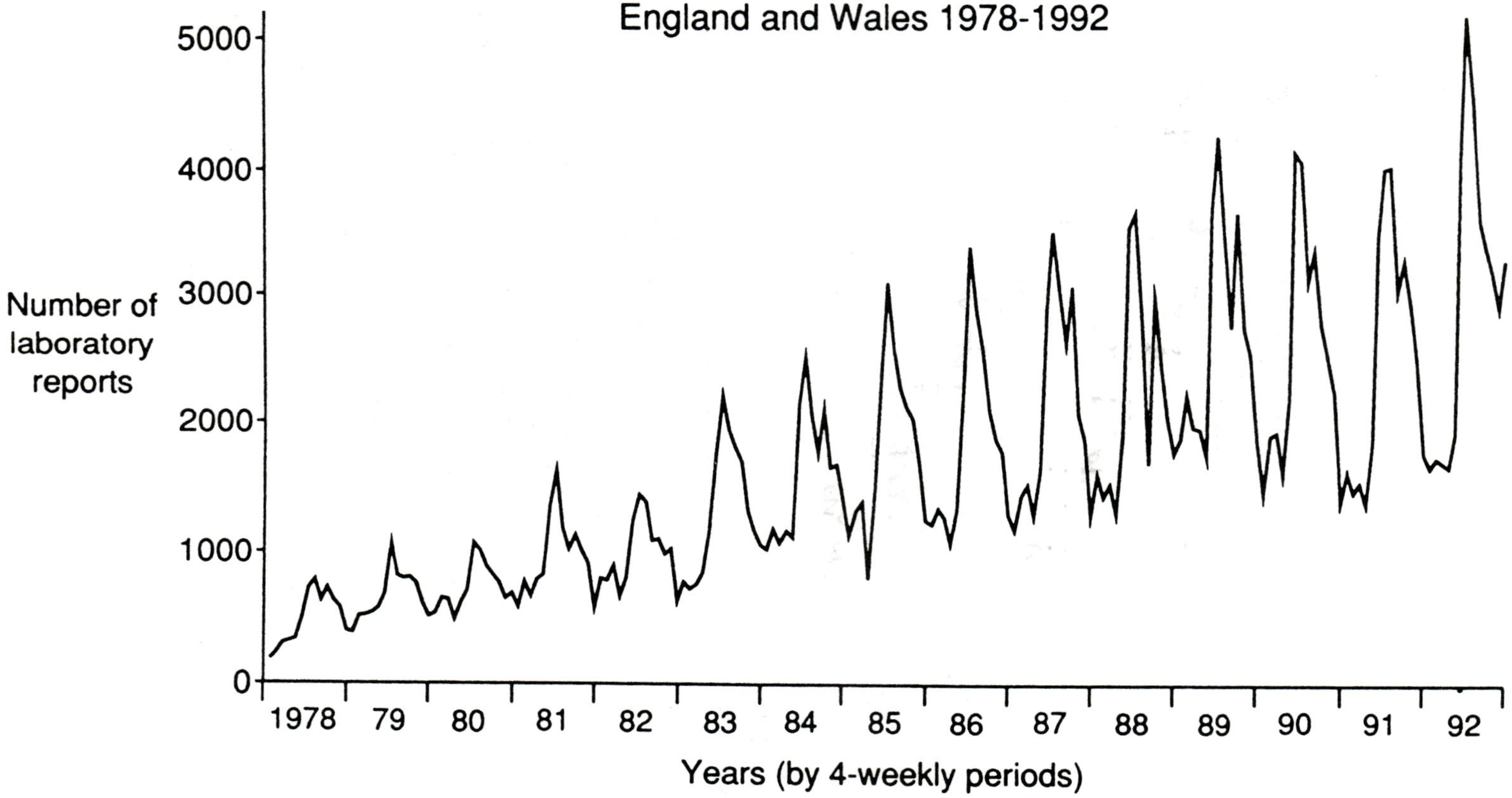

Pre 1989 figures may include some reports from Northern Ireland

Source: PHLS
Prepared by CDSC

Figure 4.5

Campylobacter Sp.

Faecal Isolates, England and Wales 1989-1991
Rates per 100,000 population by NHS Region

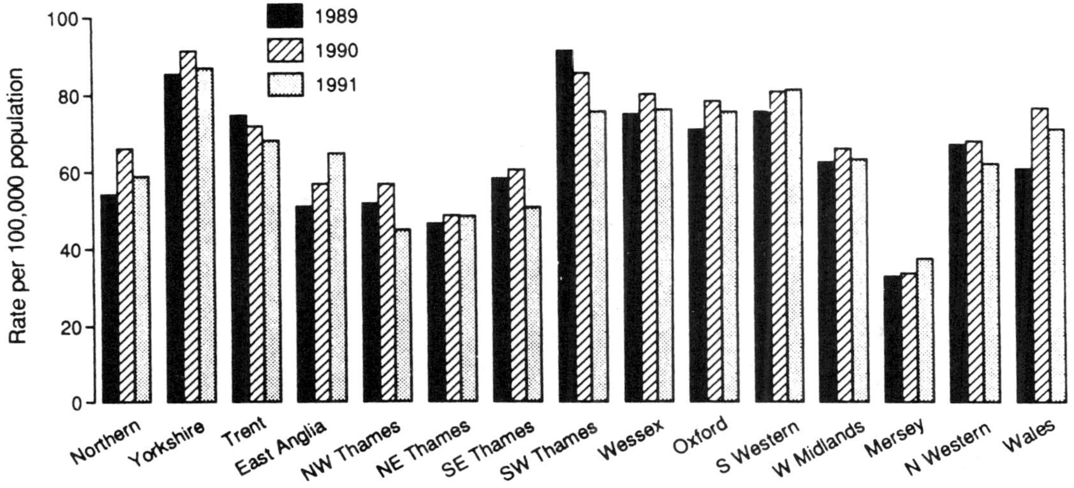

Source: PHLS
Prepared by CDSC

Conclusions

4.14 The true incidence of *Campylobacter* infection may be underestimated at the moment. Even on the basis of current knowledge, the known incidence of *Campylobacter* infection poses a major public health problem. We hope that the study recently commissioned by the Steering Group on the Microbiological Safety of Food into the incidence and aetiology of infectious intestinal disease in England will provide useful data. **(C4.1)**

4.15 Most cases of *Campylobacter* infection are apparently sporadic, and outbreaks are rarely identified. The reasons for this are unknown. **(C4.2)**

4.16 Throughout the UK *Campylobacter* mainly affects young adults. The apparently high incidence in young children is related to the disproportionately high faecal sampling rate in 1-4 year olds. Peaks occur in late spring and, in some years, in autumn. There are also regional variations. The reasons for these variations are unknown. **(C4.3)**

Recommendation

4.17 We recommend that Government funds population studies to assess the real magnitude of campylobacteriosis and further studies of transmission to understand better its seasonality. **(R4.1)**

CHAPTER 5

SOURCES AND TRANSMISSION OF INFECTION

Introduction

5.1 The control of *Campylobacter* infection will ultimately depend on a better understanding of sources, transmission routes and the way in which campylobacters interact with their human hosts. This chapter describes the current information on sources and transmission of *Campylobacter* infection.

Sources and routes of transmission of infection

5.2 Campylobacters are part of the normal intestinal flora of a wide range of wild and domestic animals and birds. Organisms may be transferred to the surface of livestock carcases and offal as a result of faecal contamination during the slaughtering process (see Chapter 6). Poultry meat is most commonly affected.[78, 79] *Campylobacter* may also be present in raw cows' milk,[80, 81] and microbiological studies have shown that *Campylobacter* is common in sewage and may be cultured from untreated water.[79] The relative importance of these various sources for human infection is unknown.

5.3 As indicated previously (see Chapter 4) outbreaks are rarely identified. Even where outbreaks are identified, in contrast to *Salmonella*, it is not always possible to identify the vehicle of infection. However, specific vehicles from outbreaks which have been investigated include undercooked meats, especially poultry meat,[82] raw milk,[81] milk that may have been inadequately pasteurised,[83] untreated water,[84] water from storage tanks which may have been contaminated after treatment,[85] and undercooked beefburgers, sliced roast beef and lamb kebabs.[86] Raw clams were considered to be the vehicle of infection in an outbreak in New Jersey USA, but the underlying cause was probably sewage pollution of the growing beds.[87] More unusual vehicles of infection which have been described in outbreaks include salads and cake icing, and are probably explained by cross-contamination.[87]

5.4 Studies of sporadic cases have implicated vehicles similar to those which have caused outbreaks. For example, the consumption of chicken, in particular undercooked chicken has been implicated,[42, 88, 89] as has the handling of raw chicken,[90] the consumption of raw milk,[91, 92] the ingestion of untreated water,[85, 93] and the consumption of barbecued meals.[79] Additional risk factors which have been identified in cases of sporadic infections include contact with pets, particularly puppies with diarrhoea[94, 95] and, in some parts of Britain in the spring, drinking milk that has been delivered in bottles to the door and has been pecked by birds. In the USA consumption of mushrooms has also been implicated.[96] However, studies of possible sources for sporadic infection fail to explain most of the cases.

Poultry

5.5 Poultry is generally recognised as an important reservoir of *Campylobacter* infection for humans. Studies of retail broiler chickens have demonstrated contamination rates of between 30% and 100%.[10, 78, 97, 98] Levels of surface contamination between 10^6 and 10^7 organisms per chicken have been recorded.[99] The contamination rate of fresh poultry is higher than that of frozen poultry.[100] It has been suggested that the increase in the reported incidence of *Campylobacter* infection over recent years may be linked to the increase in consumption in the UK of fresh (as against frozen) poultry.[101]

5.6 Evidence for the importance of poultry as a cause of human disease comes from a number of sources. In England and Wales poultry is reported as one of the most commonly suspected vehicles of infection in outbreaks. Between 1984 and 1987 119 outbreaks were reported to CDSC. Of the 82 of these in which food or water was implicated, milk was the suspected vehicle of infection in 37, poultry in 25 and water in 7.[102] Studies of sporadic infections have also implicated poultry, in particular the consumption of undercooked chicken, and the handling and preparation of raw chicken. It has also been shown that similar serotypes can be isolated from poultry on retail sale and humans with enteritis.[33, 79] In another study in the south of England control of *Campylobacter* infection in a broiler chicken farm by water chlorination and other measures led to the disappearance of the same serotype from human cases in the surrounding area.[102] Further evidence for the role of poultry in the transmission of *Campylobacter* comes from reports of antibiotic resistance in human strains after similar reports in poultry strains. A study in the Netherlands reported that between 1982 and 1989 the prevalence of quinolone resistant *Campylobacter* strains isolated from poultry products increased from 0% to 14% and during the same period the prevalence of resistance in human isolates increased from 0% to 11%.[103]

5.7 Cross-contamination from raw poultry is likely to be an important means of transmission of infection from poultry to man, but proper cooking will destroy the organisms. Experiments have shown that *C.jejuni* is easily transferred from raw chicken to cutting boards, plates and hands, and the organism has been isolated from raw vegetables and cooked chicken products which were in contact with plates on which raw chicken products had previously been placed.[104] In one study the risk of infection was inversely associated with the frequency of using soap to clean the kitchen cutting board.[96] Further, *Campylobacter* can survive for up to an hour on hands if fluids from carcases are not washed away,[105] presenting a potential route for contamination of other foods being prepared and possibly infection of the food handler. However, asymptomatic excretion of *Campylobacter* is unusual,[42, 106] and infected food handlers do not seem to present a risk. In combination with the low infective dose, cross-contamination may be the reason why studies of sporadic cases fail to explain all or even most of the cases.

5.8 It has been suggested that increased consumption of inadequately cooked barbecued chicken may in part account for the early summer peak in *Campylobacter* infections, and may also account for the higher incidence noted in young adults (see Chapter 4).

Milk

5.9 Contaminated cows' milk has long been recognised as an important vehicle/source of *Campylobacter* infection for humans. The most likely way for *Campylobacter* to contaminate the milk is as a result of poor milking parlour hygiene which can lead to high levels of faecal contamination by *C.jejuni* in raw milk.[80] *C.jejuni* can cause mastitis in cows and there is evidence that udder excretion was the cause of one milkborne outbreak.[81]

5.10 Effective pasteurisation eliminates campylobacters from milk (see Chapter 7). In Scotland, where the sale of unpasteurised cows' milk has been banned from 1 August 1983, there has been a virtual cessation of reported outbreaks of *Campylobacter* infection associated with milk.[107]

5.11 In England and Wales where the sale of unpasteurised milk is still permitted, 42 of the confirmed outbreaks of campylobacteriosis reported to CDSC in the period 1980-89 were associated with milk (see Table 5.1). Although sales of unpasteurised milk account for less than 1% of the liquid milk market, 36 (85.7%) of these 42 outbreaks were associated with unpasteurised milk. The number of people affected was small. Nevertheless, in recognition of the risks of contracting campylobacteriosis as well as salmonellosis and other foodborne infections from unpasteurised milk, in 1990 the Government introduced more explicit labelling requirements and more rigorous microbiological standards for unpasteurised milk. It is too early to assess fully the effectiveness of these new regulations in reducing campylobacteriosis associated with unpasteurised milk, but fewer outbreaks have been reported in 1991 and 1992 (see Table 5.1).

5.12 Although pasteurisation is an effective means of controlling *Campylobacter*, as Table 5.1 shows, process failures and post-pasteurisation contamination can result in outbreaks of milkborne campylobacteriosis. Pasteurisation must therefore be carefully controlled.

Table 5.1

OUTBREAKS OF *CAMPYLOBACTER* INFECTION ASSOCIATED WITH MILK AND MILK PRODUCTS CONFIRMED REPORTS TO CDSC ENGLAND AND WALES 1980-1992

YEAR	VEHICLE OF INFECTION	NUMBER OF OUTBREAKS	NUMBER AFFECTED
1980	Unpasteurised milk	5	163
	Pasteurised milk [a]	1	30
		6	193
1981	Unpasteurised milk	7	213
1982	Unpasteurised milk	3	200
	Pasteurised milk [b]	1	400
		4	600
1983	Unpasteurised milk	3	85
+ 1984	Unpasteurised milk	3	76
1985	Unpasteurised milk	6	178
1986	Unpasteurised milk	3	24
1987	Unpasteurised milk	6	332
	Pasteurised milk [c, d]	2	526
	Milk shake	1	37
		9	895
1989	Pasteurised milk [e]	1	14
++ *1990	Unpasteurised milk	1	4
++ *1991	Unpasteurised milk	1	4
++ *1992	Unpasteurised milk	1	>100
	Pasteurised milk [f]	1	>100

Source: PHLS CDSC

* Provisional or unpublished data

+ In this year one incident involving 3 people attributed to goats' milk was also recorded.

++ In each of these years a number of incidents attributed to bird-pecked bottled milk were also recorded.

[a] Milk probably contaminated when pasteurised milk was added to unsterilised containers.

[b] Probably due to mistaken addition of raw to pasteurised milk.

[c] Unsatisfactory cleaning and storage facilities for milk.

[d] Incorrect operating procedures and leaking valve at plant.

[e] Supplied from dairy as heat treated. No further information available.

[f] Pasteurisation failure.

5.13 The extent to which raw milk is a cause of sporadic cases remains uncertain. However, a comparison of data from Scotland with that from England and Wales showed that although milkborne outbreaks were controlled by banning the sale of raw milk, the same proportionate increase in sporadic cases of *Campylobacter* subsequently occurred. This was interpreted by the Richmond Committee to mean that sporadic cases in the UK are not due to undetected outbreaks of milkborne infection.[102]

5.14 More recently a number of studies have reported a link between sporadic cases of *Campylobacter* enteritis and the consumption of doorstep delivered milk from bottles that have been pecked by birds, particularly magpies and jackdaws.[108, 109, 110] This is largely confined to the height of the breeding season when the parent birds face difficulties in feeding their chicks, and coincides with the late spring/early summer peak of human infections. In some areas it is considered that this transmission route could account for almost all the increase in human cases.[110]

5.15 Intervention studies where the public have been warned of the potential risks associated with the consumption of bird-pecked milk and advised to take appropriate precautions have been disappointing. In some cases milk distributors have been reluctant to deliver advisory leaflets to consumers,[110, 111] but even where milk distributors have co-operated consumers often did not appear to follow the advice.[112] An alternative approach would be for industry and Government to consider development of better protective packaging for bottled milk, in particular the replacement of foil caps on milk bottles with a stronger material.

Water

5.16 There have been several small waterborne outbreaks of *Campylobacter* enteritis in the UK, mainly due to defective chlorination of private water supplies. No other geographical associations with specific areas of water supply were detected, suggesting that unrecognised waterborne outbreaks do not contribute significantly to the large number of sporadic cases.[102] Nevertheless *Campylobacter* has been isolated from water courses and sea water throughout England.[113] Illness has been reported in people who consumed untreated water from lakes and rivers.[93,114] In one study, the concentrations of campylobacters in surface waters were at their lowest at the same time as peak infections were occurring in the community.[115] As indicated in Chapter 2, there is evidence that *Campylobacter* can survive in water in a viable but non-culturable form, and remains infectious.[8, 9] Epidemiological evidence suggests that viable but non-culturable campylobacters may have given rise to disease during the course of an outbreak at times when water sampling found no culturable forms of the organism to be present.[85]

Red meat

5.17 Red meat might also be a source of human infection. As with poultry, cross-contamination is likely to be the most important means of transmission. Although early studies indicated that the incidence of *Campylobacter* in red meats on retail sale was very low, only 1% in the UK,[116] and 5.1% in the USA,[78] a more recent study in the UK has found higher rates of contamination.[79] In this more recent study, *Campylobacter* was found in 47% of samples of offal, 23.6% of samples of beef, 18.4% of samples of pork, and 15.5% of samples of lamb taken from meats on retail sale in the Reading area between October 1984 and September 1986. The *Campylobacter* serotypes isolated from offal, beef and lamb were very similar to those which were isolated from humans with enteritis. Most of the isolates from pork were *C.coli* of serotypes dissimilar to those isolated from humans.[79] This latter finding is similar to previous work done in Holland.[117]

Other means of transmission

5.18 Infection may also occur by direct contact with affected animals. Pets, usually dogs and/or cats with diarrhoea, have been shown to present a risk, especially to children.[94, 118]

5.19 Person to person transmission appears to be uncommon because family clusters and secondary transmission following point source outbreaks are seldom observed. This route of transmission has been reported when the index cases were young children.[119] Although the infectious dose has been shown to be low, and diarrhoeic faeces may contain as many as 10^6-10^9 organisms per gram, survival characteristics of the organisms (see Chapter 2) probably account for lack of person to person transmission.[106]

5.20 As intestinal carriage is uncommon in the UK (see Chapter 3), infected food handlers do not seem to present a major risk for transmission of infection.

National Case Control Study

5.21 Between June and December 1990 PHLS CDSC carried out a national case control study (previously piloted on a small scale in Southampton) in an attempt to elucidate possible risk factors for the acquisition of *Campylobacter* in primary, indigenous sporadic cases.

5.22 The study confirmed the previously demonstrated associations between *Campylobacter* infection and occupational exposure to raw meat, exposure to pets with diarrhoea and the ingestion of untreated water from lakes, rivers and streams. More surprisingly (and in contrast to the pilot) the study also showed that the consumption of chicken bought raw and cooked in the home was significantly associated with a decrease in risk, i.e. it was apparently protective.[120] In his report and presentation to the Committee Dr John Cowden, who led the National Case Control Study, suggested this finding may be the result of i) systematic bias in the study, which he thought was highly unlikely, and for which he could find no evidence, ii) an association between the consumption of chicken and another characteristic which was protective but undemonstrated (eg people who buy whole rather than portioned chicken have higher levels of kitchen hygiene), or iii) a genuine association. He further suggested that if the last explanation is the correct one, it was possible that immunity due to repeated exposure had led to what is otherwise a risk factor in those who are susceptible appearing to be protective overall.

5.23 The Committee commended Dr Cowden's study but together with him concluded that it did not give the information on which new conclusions could be drawn. The method of looking at risk factors was one which presented particular difficulties of interpretation in cases like *Campylobacter* where it was likely there were many vehicles of infection distributed in both time and place. The study therefore reinforced the need as set out in Chapter 2 for more widely available reliable typing schemes for *Campylobacter* to assist in the analysis of epidemiological studies as well as tracing routes of transmission. When such typing schemes are available, the value of future epidemiological studies would be greatly enhanced if investigators incorporated in their studies an assessment of the immune status of individuals to *Campylobacter*, for example by the detection of *Campylobacter* specific secretory IgA in faeces.

Conclusions

5.24 The sources and routes of transmission of *Campylobacter* infection are not yet fully understood, but there is strong circumstantial evidence to suggest one major source is by poultry, transmission being either directly through consumption of undercooked chicken or by cross-contamination of other foods in the kitchen. **(C5.1)**

5.25 Water and unpasteurised milk have been associated with outbreaks of campylobacteriosis, and one proven route of transmission for sporadic cases is consumption of doorstep delivered milk from bottles that have been pecked by birds, although this does not account for the majority of cases. **(C5.2)**

5.26 Clarification of the sources and routes of transmission would be aided if reliable sub-typing schemes were more widely available, and if epidemiological studies could take into account the immune status of individuals. **(C5.3)**

Recommendations

5.27 We have already recommended that Government fund research to develop methods of sub-typing, and to develop better detection methods for viable non-culturable forms of *Campylobacter* (see Chapter 2). This research will help the clarification of sources and routes of transmission (see Chapter 2).

5.28 We recommend that industry and Government consider the development of better protective packaging for bottled milk, in particular the replacement of foil caps on milk bottles with a stronger material. **(R5.1)**

CHAPTER 6

CAMPYLOBACTER IN ANIMALS

Introduction

6.1 The *Campylobacter* species recognised for its veterinary importance was originally named *Vibrio fetus* and was isolated from the uteri of recently aborted sheep and from the fetuses of aborted cows.[121, 122] The major human pathogens, *C.jejuni* and *C.coli*, are associated with enteric disease in young cattle, pigs, dogs and cats. Adult animals often carry *C.jejuni* and *C.coli* asymptomatically, as do poultry.

6.2 The mechanisms by which campylobacters cause disease are discussed in Chapter 3 and detailed in Appendix 3. However, it is difficult to determine whether the apparent differences in pathogenicity between animals, birds and man are due to host factors or bacterial factors.

Poultry

6.3 Thermophilic campylobacters (see Chapter 2), such as *C.jejuni* and *C.coli*, are found in the intestinal tracts of a wide variety of birds, including domestic poultry. There is little evidence that they cause disease, existing instead in a balanced relationship with the bird as commensals, i.e. the carrier state is the norm.

6.4 The origin of infection and mode of transmission within commercial poultry flocks is uncertain. Newly hatched chicks appear to be uninfected, the onset of colonisation occurs when the birds are 3-5 weeks of age.[123, 124] Under experimental conditions, however, this lag phase is not seen as chicks are susceptible to as few as 40 colony forming units of a laboratory strain when one day old.[125, 126] Rigorous hygiene and competitive exclusion can extend this lag phase significantly, suggesting it might be possible to prevent or delay colonisation during the limited life of broiler chickens. The finding that eggs from infected breeding flocks are free of infection suggests that vertical transmission from breeding flocks via hatcheries is unlikely.[125, 127, 128]

6.5 Contamination from the environment has been suggested as the most likely source of infection for commercial broiler rearing houses, with transmission possibly occurring via the footwear and clothing of farm personnel. However, it has been shown that *C.jejuni* can survive for extended periods in broiler houses and this may mean that chicks acquire infection from the environment if rearing houses are not effectively cleaned and disinfected between flocks.[129]

6.6 Water might also act as a vehicle of infection as it does in humans (see Chapter 5), and one study has shown that the prevalence of infection of *C.jejuni* can be reduced by chlorinating water supplies to poultry houses.[10] Feed is not thought likely to be an important source of infection as the moisture content of poultry feed is only 8-10% and *C.jejuni* is susceptible to drying.

6.7 The prevalence of *Campylobacter* infection in UK flocks is not well established and temporal, environmental and host factors have been little investigated. Estimates of broiler flock prevalence in other countries range from 85% in Canada, 50% in Sweden to 12% in Switzerland.[123, 130, 131] A PHLS study has shown that at slaughter, 76% of samples from broiler, broiler breeder and laying flocks collected throughout South West England were *Campylobacter* positive.[132]

Poultry processing

6.8 Poultry slaughtering is a multi-stage operation and modern plants with their emphasis on high throughput rates and automation provide ample opportunities for the cross-contamination of carcases via soiled equipment or water. Studies have shown that campylobacters can survive various processing operations and cross-contamination can occur at a number of points on the slaughter line. These are notably via scalding tanks, plucking machines and automated evisceration equipment where damage to the internal organs may occur resulting in contamination of the carcase with gut contents.[133]

6.9 After evisceration and final washing, carcases are chilled either by immersion in water, blast chilling with cold air or spray chilling. The former is used primarily for birds that are to be frozen and the latter for birds to be sold fresh. Immersion chilling can increase cross-contamination, which could be reduced by the use of chlorinated water, although apparently *C.jejuni* can survive in chilling tank water containing 15ppm chlorine.[133] This method is unpopular with consumer organisations as it increases water uptake by the carcase. Air drying could reduce *Campylobacter* contamination but may be inefficient because of the surface and texture of poultry carcases. In a study to determine the influence of different processing technologies on the contamination of chicken carcases for retail sale, immersion chilling allowed about 60% of carcases to be contaminated, whereas air chilling reduced the contamination rate to approximately 20%. *C.jejuni* has been shown to remain viable on carcases stored at -20^{0}C for three months and at 4^{0}C for seven days.[134, 135] Studies of retail broiler chickens have detected contamination rates between 30-100% (see Chapter 5).

Other agricultural livestock

6.10 *C.jejuni* is commonly present in the bovine intestinal tract,[80] and has been implicated as a possible cause of diarrhoea in calves. There may be a seasonal pattern to *C.jejuni* carriage by cattle but this has not been thoroughly investigated.[42, 136] Studies have indicated that contaminated water may be an important source of infection. Herds with access to rivers or streams when grazing have been shown to have a greater incidence of infection than those drinking only mains water.[80] *C. fetus* subspecies *venerealis* causes infertility and early embryonic death in cattle and may cause abortion in a small percentage of infected cows.

6.11 *C.fetus* subspecies *fetus* and *C.jejuni* cause abortion in sheep.[137] The disease in its sporadic or enzootic form is well known throughout the world and heavy losses can occur during the lambing season. There is little evidence that *C.jejuni* causes enteric disease in sheep.[138]

6.12 Porcine proliferative enteritis is an important infectious disease of pigs and is associated with intracellular campylobacter-like organisms (CLO).[139] CLO have not yet been cultured by conventional bacteriological techniques. Currently there is no evidence of pathogenicity in humans. *C.coli* causes diarrhoea in piglets while *C.jejuni* less commonly causes enteric disease in older pigs. Both *C.jejuni* and *C.coli* are also present in asymptomatic animals.

Red meat slaughterhouses

6.13 Campylobacters are present in the gastrointestinal tract of clinically normal red meat animals, but red meats have been shown to yield fewer isolates of *C.jejuni* than poultry.[78, 79] The reasons for this could be due to differences in the initial level of infection on the farm. However, differences in the operation of red meat slaughterhouses as compared to those of poultry, for example the low throughput rate, means that the possibility of any cross-contamination is reduced. Furthermore, the practice of rodding and bunging, which involves sealing the oesophagus and anus in cattle and sheep, can reduce the risk of cross contamination during the carcase dressing process.

6.14 The handling of red meat carcases after slaughter, which includes hanging and chilling, causes the carcase surface to dry out, and results in a reduction in the number of campylobacters present due to the sensitivity of the organism to desiccation. Offals appear to be more contaminated with campylobacters than meat, with one study showing that 30% of sheep and 10% of cattle offals in abattoir and retail stores were affected.[140] A more recent study has shown that 47% of samples of offal on retail sale were contaminated (see Chapter 5).[79]

Pet animals

6.15 *C.jejuni* can be isolated from both healthy and diarrhoeic dogs.[141] Several studies have shown no significant difference between prevalence in normal adult animals and those suffering from diarrhoea. Reported prevalences vary considerably from a few per cent to more than 50%.[142] An increase in prevalence of infection is frequently associated with the presence of other pathogens such as parvovirus, *Salmonella, Giardia* and other parasites. Stress such as surgery, pregnancy and other illnesses may also increase excretion rates. Isolation rates are greater in puppies than in mature dogs and in kennel populations than in household dogs.

6.16 *Campylobacter* infection in cats and other pet animals is less well described, but it has been reported in up to 45% of non-diarrhoeic cats kept in one RSPCA premise, although in domestic cats housed individually the infection rate appears to be low.[141] Kittens appear to have a greater prevalence of infection than adult cats, although this has been disputed.

Conclusions

6.17 Campylobacters are not major veterinary pathogens. They are found in the gastrointestinal tract of a wide variety of animals and birds without causing disease. *C.fetus* is a cause of infertility in cattle and abortion in sheep, and *C.jejuni* and *C.coli* cause enteric disease in cattle and pigs. **(C6.1)**

6.18 *C.jejuni* has been isolated from healthy dogs and dogs with diarrhoea especially puppies. Infection in cats and other pet animals is less often detected. **(C6.2)**

6.19 The origin of infection and mode of transmission for commercial poultry flocks is uncertain but may be via contaminated water or the environment. **(C6.3)**

6.20 Red meats yield fewer numbers of campylobacters in comparison to poultry. **(C6.4)**

6.21 Cross-contamination of carcases may occur during slaughtering and processing, but the automation and high throughput of modern poultry plants makes this more likely with poultry than with red meat. **(C6.5)**

Recommendations

6.22 We recommend that Government should fund research to establish:

- why some campylobacters cause disease in humans and in some animals but not in other animals; and,

- the prevalence of *Campylobacter* infection in UK poultry flocks, the origins of infection and the routes of transmission, and the mechanisms by which infection may be controlled. **(R6.1)**

CHAPTER 7

CAMPYLOBACTER INFECTIONS IN HUMANS : POSSIBILITIES FOR PREVENTION

Introduction

7.1 At the present time campylobacteriosis is generally considered to be a foodborne infection for which foods of animal origin constitute the most important source. Control and prevention therefore ultimately depend on reducing the numbers of campylobacters in the whole food chain. The fact that *Campylobacter* organisms are commonly found in the intestines of a wide range of wild and domestic animals, combined with the cross contamination which occurs using current methods of slaughtering and processing of meat for human consumption, means that reduction of the number of micro-organisms in the whole food chain must necessarily be a long-term goal rather than something which will be achieved rapidly.

7.2 Achievement of this goal will be greatly helped by an increase in our understanding of the physiology and pathogenicity of *Campylobacter*. We have made specific recommendations for research on these and other topics in earlier chapters. In particular we believe it is important to develop sub-typing mechanisms to clarify sources and routes of transmission (see Chapter 2), as this will provide the opportunity to develop specific preventive measures.

7.3 In the shorter term a more practical approach is to consider the control of *Campylobacter* in foods based on an understanding of their growth and survival characteristics. Such an approach is used in the Hazard Analysis and Critical Control Point (HACCP) system. We believe this must also be combined with general food hygiene training and education for food handlers and the public.

7.4 In this Chapter we consider the growth and survival characteristics of *Campylobacter* in food to form the basis for recommending practical measures that can be taken now by industry and consumers to prevent *Campylobacter* infection. We also set out our recommendations for food hygiene training and education.

Growth and survival characteristics of *Campylobacter* in food

Storage Temperature

7.5 Although *C.jejuni* is not reported to grow in foods at temperatures below 30°C, it is capable of growth in foods above this temperature.[143, 144, 145, 146, 147] There are also a number of reports of its survival in foods held at low temperatures, for example, *Campylobacter* has been shown to survive well in beef and ground beef held at 1°C and 10°C with a reduction in numbers of less than 10% in 48 hours.[148] Campylobacters are reported to survive less well in foods stored at room temperature.[143, 149]

7.6 Freezing has been reported to reduce the level of, but not eliminate, *C.jejuni* in foods. *C.jejuni* can remain viable on naturally contaminated chicken carcases stored at -20°C for 7 days.[135]

Heating

7.7 *C.jejuni* is a non-sporing organism so it will not survive heat treatments sufficient to eliminate its vegetative cells (see Chapter 2). The MAFF Micromodel is a predictive micromodel that can provide detailed estimates of the survival of *C. jejuni* under various conditions. The time taken at a particular temperature to reduce the number of vegetative cells to 10% of the initial number is known as the D-value. Further information provided by MAFF Micromodel on how D-values of *C.jejuni* vary with changes in temperature, pH and salt can be found in Appendix 1.

7.8 Cooking procedures sufficient to kill *Salmonella* species will also kill *C.jejuni*, nor will *C.jejuni* survive the minimum pasteurisation treatment legally required for drinking milk and certain milk products,[150, 151] but care must be taken to avoid post-pasteurisation contamination (see Chapter 5).

7.9 *C.jejuni* is also eliminated by heat treatments used in meat processing.[15] For example, significant numbers of *C.jejuni* in raw ground beef were reduced to undetectable levels within 10 minutes when meatballs were cooked to an internal temperature of 60°C.[152] *C.jejuni* will also easily be eliminated by the heat treatment used in the processing of canned and other similarly processed shelf stable foods.

Salt (Sodium Chloride)

7.10 *C.jejuni* is not eliminated by the salt concentration commonly found in most foods. For example, at 42°C some strains are able to grow in the presence of 1.5% but not 2% salt. At 4°C growth is inhibited by 0.5% salt, but viable cells can still be recovered after 14 days.[153]

Water activity (A_w)

7.11 Water activity is a measure of the water in a food that is available for microbial growth.[154] In practice, the effect of this is similar to that of salt: drying inhibits growth of the organism but does not eliminate it entirely. *C.jejuni* is sensitive to drying at room temperature, but under certain refrigerated conditions the organism can remain viable for several weeks.[15]

7.12 When considering water activity it is important to distinguish between its effect on *C.jejuni* spread throughout the foodstuff compared to its effect at the surface of the food, e.g. on the skin at the surface of an animal carcase (see Chapter 6). Although it was possible to recover campylobacters from 26% of moist areas of pork carcases sampled, they could only be recovered from 2% of dry areas.

Acidity/alkalinity (pH)

7.13 *C.jejuni* grows best in laboratory media in the pH range 6.5-7.5. It does not grow below a pH of 4.9. On meat, growth of *C.jejuni* does not occur at 37°C when the pH is normal (5.8), but does when the pH is higher (6.4).[144] Below a pH of 4.9 in laboratory media *C.jejuni* is inactivated at a rate which is dependent on temperature. Inactivation is most rapid at 42°C, intermediate at 25°C, and slowest at 4°C.[156] On the surface of chicken broiler halves, lactic and acetic acids have been demonstrated to reduce the numbers of *C.jejuni*,[15] but there are no reports that the pH levels commonly found in foods will totally eliminate the organism.

Atmosphere

7.14 Several studies have shown that survival of *C.jejuni* on foods may vary depending on the composition of the atmosphere in which the food is stored.[157] For example, one recent study compared the survival of *C.jejuni* on inoculated turkey roll slices in 7 different atmospheres composed of varying percentages of carbon dioxide, nitrogen and oxygen. Compared with air, atmosphere containing 40-60% carbon dioxide or 100% nitrogen allowed greatest survival.[158]

7.15 The gas mixture used in modified atmosphere packaged (MAP) foods is chosen to meet the needs of the specific food product, but for nearly all products this will be a combination of carbon dioxide, oxygen and nitrogen. More information is needed to investigate the possibility of enhanced survival of campylobacters in MAP foods, particularly in view of the increasing use of MAP.

Other factors

7.16 There is an apparent lack of information on effects of preservatives such as nitrite, nitrate, sorbate and nicin on *C. jejuni*. In keeping with the other parameters, like salt, it is more than likely that preservatives will inhibit the growth of the organism but will not eliminate it. Although the data available are limited, studies have shown that *C. jejuni* is as sensitive or more sensitive than salmonellae to gamma radiation.[159,160] If food irradiation is to be considered as a control option for *C. jejuni*, further work would be needed to examine the effect of irradiation on other species and strains, and on a wide range of substrates and conditions. The potential for recovery of *Campylobacter* following irradiation would also need further examination.

Hazard Analysis and Critical Control Point (HACCP) analysis

7.17 HACCP uses the occurrence, growth and survival characteristics of organisms in food to develop preventive measures, by providing a structured approach which can be used to analyse the potential hazards in an operation and to decide which are critical to consumer safety. Critical Control Points (CCPs) are monitored and remedial action is taken if conditions at the CCP are outside safe limits.[161, 162] In the UK a

HACCP based approach to the control of hazards in foods has been promoted by Government and adopted by many sectors of the industry.

7.18 One of the strengths of the HACCP approach is that it can be applied throughout the whole food chain, from the purchase of raw materials to the serving of the food for final consumption. Although there are a number of sectors in the food chain, there are only a few types of operation involved: purchase, storage, cooking and cooling. The preventive measures that need to be taken during these operations to eliminate or reduce *Campylobacter* apply equally to businesses and consumers, and each party has an equal responsibility to ensure that they are implemented appropriately.

Purchase

7.19 The purchase of raw materials is the first step in the food chain and as such is critical. If campylobacters are present on any raw material then they must be eliminated at some point by a processing step, otherwise they may present a hazard when food is consumed.

7.20 Campylobacters can be found in raw materials of animal origin, especially in poultry. Only canned materials or those that can be demonstrated to have had an adequate heat treatment while in their final packaging can be considered to be free from the organism. All other raw materials of animal origin should be considered potentially contaminated.

7.21 The typical control options open to the industry at this stage are for buyers to undertake hygiene inspections of the supplier's premises and raw material specifications, which will help ensure the quality of the raw material.

7.22 Consumers should, in addition, follow the advice given in the "Shopping for Food" section of the Foodsense Food Safety Leaflet.[163] This includes avoiding shops where cooked and raw meat are not separated and where assistants are seen to handle raw food and then touch cooked food without washing their hands thoroughly in between. Consumers should inform their local Environmental Health Department when these unacceptable hygiene practices are encountered.

Storage

7.23 *Campylobacter* is unlikely to multiply in frozen or chilled foods or foods stored at room temperature. As outlined earlier in this Chapter, *Campylobacter* is likely to survive in these conditions. For the control of other pathogens and spoilage organisms, producers and retailers should store at the temperatures stated in the Food Hygiene (Amendment) Regulations 1990[164] and 1991.[165] Consumers should store food at the temperature stated on the label as far as is practicable and to the extent the design of domestic refrigerators allows. Recent research has indicated that temperature control in some domestic refrigerators is variable.[166]

7.24 Because campylobacters will not grow in foods exposed to normal atmospheres, the storage time is not relevant to the practical control of this organism in most foods.

For the control of other organisms and spoilage organisms, foods should be consumed before the "use by" date stated on the label.

Cooking

7.25 Heating is the only operation that can be controlled to eliminate the organism consistently, and as such is one of the CCPs in the food chain. We endorse the Government's recommendation of cooking foods to a minimum internal temperature of 70°C for 2 minutes or equivalent, [163,167] as this will more than adequately eliminate campylobacters (as well as other vegetative pathogens). Equivalent heat treatments are shown in Table 7.1.

7.26 It is also important that heating equipment, including microwave ovens, must be capable of consistently achieving this time/temperature combination in every part of the food. The cooking process must be monitored and remedial action taken if the time/temperature combination is not achieved. Further details are given in the leaflet "Safer Cooked Meat Production Guidelines",[167] and for consumers in the "Cooking Food" section of the Food Safety Leaflet.[163]

7.27 Furthermore, industry needs to ensure that milk and milk products comply with the pasteurisation requirements laid down in appropriate UK and EC legislation. Guidelines for the hygienic pasteurisation of dairy based products are available.[168, 170]

Cooling

7.28 After heating, food must spend an absolute minimum of time in the temperature zone 30°C-45°C i.e. it must be cooled immediately to prevent the potential growth of *Campylobacter*.

Prevention of cross contamination

7.29 Because such low numbers of *Campylobacter* can cause disease symptoms (see Chapter 3), it is highly likely that cross contamination might place enough organisms on a previously uncontaminated food to cause infection. Therefore, steps need to be taken to avoid the risks of cross contamination occurring.

7.30 Although there are many ways in which cross contamination can occur, they all fall into two broad categories : direct and indirect. Direct cross contamination is where the source of the contamination directly contacts the food e.g. a piece of raw meat touching a ready to eat food. Indirect cross contamination is where the transfer of micro-organisms is via another material such as a knife or a dishcloth. Indirect cross contamination can sometimes occur via a chain of factors, such as raw meat to chopping board, chopping board to dishcloth, dishcloth to plate, plate to food. The combinations are myriad and great care is needed to prevent it.

TABLE 7.1

EQUIVALENT HEAT TREATMENTS

Temperature °C	Time
60	45 minutes
65	10 minutes
70	2 minutes
75	30 seconds
80	6 seconds

Source: Reproduced from "Safer Cooked Meat Production Guidelines".[167]

7.31 General advice on avoiding cross contamination is available from a number of sources such as "Safer Cooked Meat Production Guidelines",[167] "Guidelines on Cook Chill and Cook Freeze Catering Systems",[170] and for consumers "Food Safety".[163] Common guidance for industry, much of which applies equally to the avoidance of cross contamination in the domestic kitchen, includes:

- raw foods must be handled in a separate area from cooked foods unless their processing is separated by time and the area effectively cleaned in between;

- hands should be washed before any food is handled, and especially after handling raw meat;

- separate utensils, including thermometer probes, should be used for cooked and raw foods (the same utensils could be used if effectively cleaned in between);

- all cleaning equipment including dishcloths and kitchen towels should be sanitised thoroughly before using in areas in which cooked foods will be handled;

- raw foods, used utensils, or surfaces likely to cause contamination should never be allowed to come into contact with cooked foods;

- where possible separate staff should be used in high risk (care) and low risk (care) areas;

- use separate refrigerators for cooked and raw foods. If this is not possible do not store raw foods above cooked foods; and,

- use only potable water for processing foods and for cleaning.

The need for food hygiene training and education

7.32 It is unlikely that the numbers of campylobacters entering the food chain will ever be reduced to zero. This means that possible sources of infection will continue to be present in the commercial and domestic kitchen. *Campylobacter* infections could be significantly reduced if there was better understanding of the need to avoid cross contamination and to cook food properly in order to kill any organisms which may be present. The public are not generally aware of the micro-organisms which most commonly cause foodborne disease unless, like *Salmonella*, they have been reported on extensively in the media. However, we are encouraged by reports that consumers and producers are increasingly attaching more importance to the need for safe food.

Broadly, the main messages on food safety which we would offer to the two primary groups of people involved are:

i. <u>for commercial food handlers</u>: to focus on training in safe handling methods;

ii. <u>for consumers</u>: education programmes should concentrate on good practices, such as speedy transport, proper storage and the hygienic preparation of foods.

7.33 The Government announced on 17 December 1992 that it would be consulting in 1993 on draft regulations to implement the EC Directive on the Hygiene of Foodstuffs which was then under negotiation. In keeping with the approach adopted in the Directive, these regulations would include a broad requirement for food businesses to supervise, instruct and/or train their staff in food hygiene in a way commensurate with their tasks in the preparation or handling of food. Following adoption of the Hygiene Directive on 14 June 1993 we understand this consultation will now take place at the end of the year.

7.34 We are also aware that a large volume of food safety information is available to the public, much of which has been provided by the major supermarket chains and the Health Education Authority in the form of leaflets. The Government has also produced a series of "Foodsense" booklets which were launched in August 1991. Avoidance of cross contamination is of particular importance in preventing *Campylobacter* infection. Particular attention should therefore be directed to this in developing further advice to consumers. We would also support improved provision of educational material for those in further/higher education, or those who have just left education, as *Campylobacter* infection peaks in young adults.

Committee's further action

7.35 We believe that measures to reduce the number of campylobacters in the food chain are likely to have a concomitant effect in reducing the load of other micro-organisms such as *Salmonella*. This points to a need for guidance to particular industries on how to reduce the numbers of micro-organisms in the food chain.

7.36 In view of the importance of poultry in *Campylobacter* and in *Salmonella* infections we believe that the first industry for which guidance should be developed is the poultry industry. We have therefore set up a Poultry Working Group which will help us to formulate advice to Ministers and industry on action which could usefully be taken to reduce the incidence of foodborne pathogens in the poultry industry.

7.37 This Working Group first met early in 1993 and intends to assess the significance of every step in the poultry production process, through four main areas:

i. the influence of breeding stock;

ii. factors in the environment and the influence of poultry house design;

iii. the spread of *Salmonella* and *Campylobacter* in poultry houses and during transport; and,

iv. slaughter house operations, further processing and packing.

Conclusions

7.38 Although *Campylobacter* does not grow in frozen or chilled foods or foods stored at normal room temperatures, it is likely to survive in these foods. **(C7.1)**

7.39 *Campylobacter* will be eliminated by the heat treatments used in the processing of pasteurised milk, ready to eat cooked meats and canned and other shelf stable foods. In the domestic kitchen *C.jejuni* will not survive normal cooking procedures. **(C7.2)**

7.40 In practice no food contains sufficient salt to inhibit growth of *Campylobacter* but the organism is likely to survive, albeit at reduced levels, in the salt concentrations found in most foods. **(C7.3)**

7.41 Drying of some foods helps to inhibit growth of *C.jejuni* but the organism is likely to survive at water activity levels found in most foods, albeit at reduced levels. Drying of the surface of a number of foods can reduce or eliminate the organism. **(C7.4)**

7.42 Acidity below pH 4.9 will effectively inhibit growth and indeed reduce the levels of *C.jejuni* in foods but has not been demonstrated to eliminate the organism totally. **(C7.5)**

7.43 The atmosphere in which food is stored may affect the survival of *Campylobacter*, but more information is needed to assess this effect and its practical implications. **(C7.6)**

7.44 *C. jejuni* is sensitive to gamma radiation, but further work is needed to examine the effects on other species and strains and in different environmental conditions. **(C7.7)**

7.45 HACCP is a useful way for industry to identify and thus control potential hazards in an operation which are critical to consumer safety. **(C7.8)**

7.46 Better design of domestic refrigerators would enable more effective chilled storage of foods in the home. **(C7.9)**

7.47 Heating is the only operation that can be controlled to eliminate *Campylobacter* consistently. **(C7.10)**

7.48 Prevention of cross contamination, particularly during the handling of raw materials of animal origin, especially poultry, is an important means of controlling transmission of *Campylobacter* to humans. **(C7.11)**

7.49 There is a need for training for commercial food handlers to focus on safe handling methods and for educational programmes for consumers to concentrate on good practices such as speedy transport, proper storage and the hygienic preparation of foods. **(C7.12)**

Recommendations

7.50 We recommend that industry should make full use of the information on survival of *Campylobacter* under various conditions of salt, temperature and pH contained in the MAFF Micromodel. **(R7.1)**

7.51 We recommend that MAFF extends the Micromodel database to include the effects of different atmospheres and preservatives. **(R7.2)**

7.52 We recommend that industry funds research to investigate the effect of preservatives and the use of modified atmospheres in processing and packaging on the survival or inactivation of campylobacters in food. **(R7.3)**

7.53 We recommend that if industry wishes to use irradiation as a means of controlling *Campylobacter*, they should build on previous research which has demonstrated that *C.jejuni* is sensitive to irradiation, and fund further research to examine the effect of irradiation on other species associated with human disease. **(R7.4)**

7.54 We recommend that all sectors of the food industry adopt a HACCP-based approach to the control of potential microbiological hazards. **(R7.5)**

7.55 We recommend that industry should consider improvements in the design of domestic refrigerators to enable consumers to maintain effective temperature control at the domestic end of the chill chain. **(R7.6)**

7.56 We recommend that industry ensures that heating equipment, including microwave ovens, are capable of consistently achieving appropriate time/temperature combinations to kill micro-organisms in every part of the food. **(R7.7)**

7.57 We recommend that consumers cook foods thoroughly, following reliable guidance, such as that given in the "Cooking Food" section of the "Food Safety" leaflet.[163] **(R7.8)**

7.58 We recommend that industry ensures that the pasteurisation of milk and milk products is carefully controlled, and that post-pasteurisation contamination is avoided. **(R7.9)**

7.59 We recommend that industry and consumers take steps to avoid cross contamination, particularly during the handling of raw materials of animal origin, by following reliable guidance such as that given in "Safer Cooked Meat Production Guidelines",[167] "Guidelines on Cook Chill and Cook Freeze Catering Systems",[170] and for consumers, "Food Safety".[163] **(R7.10)**

7.60 We recommend that the Government should introduce a statutory requirement for training of personnel involved in the production, distribution and retailing of food, and that enforcement officers should ensure that training programmes are commensurate with the risks involved in individual food businesses. **(R7.11)**

7.61 We recommend that the Government's public information programme, as it is developed, should pay particular attention to the avoidance of cross contamination as a means of preventing *Campylobacter* infection. **(R7.12)**

CHAPTER 8

CONCLUSIONS AND RECOMMENDATIONS

CHAPTER 2
THE INFECTIVE AGENT

Conclusions

C2.1 The significant microbiological properties of most *Campylobacter* species are that they do not grow at normal room temperatures or at refrigeration temperatures, neither do they grow in air. Campylobacters are sensitive to drying, and to acidic conditions. Salt levels over 2% inhibit growth, and the organisms are inactivated at temperatures of 48°C and over. Campylobacters have been shown to exist in a viable but non-culturable form, in which they are not isolated on standard culture media. It is suggested that such forms are a response to certain environmental conditions, but it is not clear whether they are infectious for humans. (2.16)

C2.2 There is a need to be able to differentiate between *Campylobacter* species and sub-types to enable better identification of sources of infection. Differentiation of *Campylobacter* species is not currently done by laboratories routinely, so the full significance of different species or sub-types in causing human disease is unknown. (2.17)

C2.3 Various sub-typing methods have been used, and new methods are being developed, but some strains are either untypable or indistinguishable using current methods. Other strains can only be accurately identified using a combination of available methods. As there is currently no central *Campylobacter* Reference Laboratory, this work lacks a focal point. (2.18)

Recommendations

R2.1 A central UK *Campylobacter* Reference Laboratory should be established to co-ordinate further investigation of *Campylobacter* strains, and we recommend that the Government consider how best this might be done. (2.19)

R2.2 In addition, we recommend that Government funds research to expand current knowledge of *Campylobacter* species, in particular:

- to establish isolation and identification methods that can be used by clinical laboratories for the detection of all clinically relevant *Campylobacter* species;

- to develop methods of sub-typing which will enable better epidemiological tracing of sources and transmission routes of human infection; and,

- to develop better detection methods for viable, non-culturable forms of *Campylobacter* in order to determine whether they play a role in the production of human enteritis. (2.20)

CHAPTER 3
DISEASE DESCRIPTION AND IMMUNE RESPONSE IN HUMANS

Conclusions

C3.1 Campylobacter infection can be caused by ingestion of a small number of bacterial cells. (3.19)

C3.2 Some individuals may have a degree of acquired immunity to *Campylobacter* infection, but the extent of such immunity in the UK population is unknown. (3.20)

C3.3 The disease symptoms in the UK are of an inflammatory type of campylobacteriosis, and differ from the secretory type of enteritis reported in developing countries, which may be due to infection by toxin producing strains. (3.21)

C3.4 The relative potential for different strains of *C.jejuni/coli* and the less commonly isolated species to cause disease in humans is not well understood at this time, and more information is needed. (3.22)

Recommendations

R3.1 We recommend that Government funds research to provide more information:

- to establish whether all strains of *C.jejuni/coli* from whatever source have equal disease causing potential for humans;

- to establish the disease causing potential of the more recently described species, *C.lari, C.hyointestinalis* and *C.upsaliensis*;

- to investigate the role of toxin-producing strains in the UK; and,

- to establish the level of immunity to *Campylobacter* in the UK. (3.23)

CHAPTER 4
EPIDEMIOLOGICAL SURVEILLANCE IN HUMANS

Conclusions

C4.1 The true incidence of *Campylobacter* infection may be underestimated at the moment. Even on the basis of current knowledge, the known incidence of *Campylobacter* infection poses a major public health problem. We hope that the study recently commissioned by the Steering Group on the Microbiological Safety of Food into the incidence and aetiology of infectious intestinal disease in England will provide useful data. (4.14)

C4.2 Most cases of *Campylobacter* infection are apparently sporadic, and outbreaks are rarely identified. The reasons for this are unknown. (4.15)

C4.3 Throughout the UK *Campylobacter* mainly affects young adults. The apparently high incidence in young children is related to the disproportionately high faecal sampling rate in 1-4 year olds. Peaks occur in late spring and, in some years, in autumn. There are also regional variations. The reasons for these variations are unknown. (4.16)

Recommendation

R4.1 We recommend that Government funds population studies to assess the real magnitude of campylobacteriosis, and further studies of transmission to understand better its seasonality. (4.17)

CHAPTER 5
SOURCES AND TRANSMISSION OF INFECTION

Conclusions

C5.1 The sources and routes of transmission of *Campylobacter* infection are not yet fully understood, but there is strong circumstantial evidence to suggest one major source is by poultry, transmission being either directly through consumption of undercooked chicken or by cross-contamination of other foods in the kitchen. (5.24)

C5.2 Water and unpasteurised milk have been associated with outbreaks of campylobacteriosis, and one proven route of transmission for sporadic cases is consumption of doorstep delivered milk from bottles that have been pecked by birds, although this does not account for the majority of cases. (5.25)

C5.3 Clarification of the sources and routes of transmission would be aided if reliable sub-typing schemes were more widely available, and if epidemiological studies could take into account the immune status of individuals. (5.26)

Recommendations

R5.1 We recommend that industry and Government consider the development of better protective packaging for bottled milk, in particular the replacement of foil caps on milk bottles with a stronger material. (5.28)

CHAPTER 6
CAMPYLOBACTER IN ANIMALS

Conclusions

C6.1 Campylobacters are not major veterinary pathogens. They are found in the gastrointestinal tract of a wide variety of animals and birds without causing disease. *C.fetus* is a cause of infertility in cattle and abortion in sheep, and *C.jejuni* and *C.coli* cause enteric disease in cattle and pigs. (6.17)

C6.2 *C.jejuni* has been isolated from healthy dogs and dogs with diarrhoea especially puppies. Infection in cats and other pet animals is less often detected. (6.18)

C6.3 The origin of infection and mode of transmission for commercial poultry flocks is uncertain but may be via contaminated water or the environment. (6.19)

C6.4 Red meats yield fewer numbers of campylobacters in comparison to poultry. (6.20)

C6.5 Cross-contamination of carcases may occur during slaughtering and processing, but the automation and high throughput of modern poultry plants makes this more likely with poultry than with red meat. (6.21)

Recommendations

R6.1 We recommend that Government should fund research to establish:

- why some campylobacters cause disease in humans and in some animals but not in other animals; and,

- the prevalence of *Campylobacter* infection in UK poultry flocks, the origins of infection and the routes of transmission, and the mechanisms by which infection may be controlled. (6.22)

CHAPTER 7
CAMPYLOBACTER INFECTIONS IN HUMANS: POSSIBILITIES FOR PREVENTION

Conclusions

C7.1 Although *Campylobacter* does not grow in frozen or chilled foods or foods stored at normal room temperatures, it is likely to survive in these foods. (7.38)

C7.2 *Campylobacter* will be eliminated by the heat treatments used in the processing of pasteurised milk, ready to eat cooked meats and canned and other shelf stable foods. In the domestic kitchen *C.jejuni* will not survive normal cooking procedures. (7.39)

C7.3 In practice no food contains sufficient salt to inhibit growth of *Campylobacter* but the organism is likely to survive, albeit at reduced levels, in the salt concentrations found in most foods. (7.40)

C7.4 Drying of some foods helps to inhibit growth of *C.jejuni* but the organism is likely to survive at water activity levels found in most foods, albeit at reduced levels. Drying of the surface of a number of foods can reduce or eliminate the organism. (7.41)

C7.5 Acidity below pH 4.9 will effectively inhibit growth and indeed reduce the levels of *C.jejuni* in foods but has not been demonstrated to eliminate the organism totally. (7.42)

C7.6 The atmosphere in which food is stored may affect the survival of *Campylobacter*, but more information is needed to assess this effect and its practical implications. (7.43)

C7.7 *C. jejuni* is sensitive to gamma radiation, but further work is needed to examine the effects on other species and strains and in different environmental conditions. (7.44)

C7.8 HACCP is a useful way for industry to identify and thus control potential hazards in an operation which are critical to consumer safety. (7.45)

C7.9 Better design of domestic refrigerators would enable more effective chilled storage of foods in the home. (7.46)

C7.10 Heating is the only operation that can be controlled to eliminate *Campylobacter* consistently. (7.47)

C7.11 Prevention of cross contamination, particularly during the handling of raw materials of animal origin, especially poultry, is an important means of controlling transmission of *Campylobacter* to humans. (7.48)

C7.12 There is a need for training for commercial food handlers to focus on safe handling methods and for educational programmes for consumers to concentrate on good practices such as speedy transport, proper storage and the hygienic preparation of foods. (7.49)

Recommendations

R7.1 We recommend that industry should make full use of the information on survival of *Campylobacter* under various conditions of salt, temperature and pH contained in the MAFF Micromodel. (7.50)

R7.2 We recommend that MAFF extends the Micromodel database to include the effects of different atmospheres and preservatives. (7.51)

R7.3 We recommend that industry funds research to investigate the effect of preservatives and the use of modified atmospheres in processing and packaging on the survival or inactivation of campylobacters in food. (7.52)

R7.4 We recommend that if industry wishes to use irradiation as a means of controlling *Campylobacter*, they should build on previous research which has demonstrated that *C.jejuni* is sensitive to irradiation, and fund further research to examine the effect of irradiation on other species associated with human disease. (7.53)

R7.5 We recommend that all sectors of the food industry adopt a HACCP-based approach to the control of potential microbiological hazards. (7.54)

R7.6 We recommend that industry should consider improvements in the design of domestic refrigerators to enable consumers to maintain effective temperature control at the domestic end of the chill chain. (7.55)

R7.7 We recommend that industry ensures that heating equipment, including microwave ovens, are capable of consistently achieving appropriate time/temperature combinations to kill micro-organisms in every part of the food. (7.56)

R7.8 We recommend that consumers cook foods thoroughly, following reliable guidance, such as that given in the "Cooking Food" section of the "Food Safety" leaflet.[163] (7.57)

R7.9 We recommend that industry ensures that the pasteurisation of milk and milk products is carefully controlled, and that post-pasteurisation contamination is avoided. (7.58)

R7.10 We recommend that industry and consumers take steps to avoid cross contamination, particularly during the handling of raw materials of animal origin, by following reliable guidance such as that given in "Safer Cooked Meat Production Guidelines",[167] "Guidelines on Cook Chill and Cook Freeze Catering Systems",[170] and for consumers, "Food Safety".[163] (7.59)

R7.11 We recommend that the Government should introduce a statutory requirement for training of personnel involved in the production, distribution and retailing of food, and that enforcement officers should ensure that training programmes are commensurate with the risks involved in individual food businesses. (7.60)

R7.12 We recommend that the Government's public information programme, as it is developed, should pay particular attention to the avoidance of cross contamination as a means of preventing *Campylobacter* infection. (7.61)

GLOSSARY

This glossary is intended as an aid to the reading of the main text and is not intended to be definitive.

ACID BARRIER OF THE STOMACH
The stomach is an acid environment and some micro-organisms will not survive the level of acidity. Thus the stomach can act as a barrier preventing some micro-organisms from passing any further down the intestines.

ANTIBODY
A protein formed in direct response to the introduction into an individual of an antigen. Antibodies can combine with their specific antigens e.g. to neutralise toxins or destroy bacteria.

ANTIGEN
A substance which elicits an immune response when introduced into an individual.

ASYMPTOMATIC CARRIER
A person infected with a micro-organism who does not suffer any resulting symptoms or disease.

BACTERAEMIA
The presence of bacteria in the blood.

BACTERICIDAL
Able to kill at least some types of bacteria.

BACTERIOPHAGE TYPING
A method for distinguishing varieties of bacteria within a particular species on the basis of their susceptibilities to a range of bacteriophages (bacterial viruses).

BACTERIUM
A microscopic organism with a rigid cell wall; often unicellular and multiplying by splitting in two.

BIOTYPING
A method for distinguishing varieties of bacteria by metabolic and/or physiological properties.

BLOOD CULTURE
A procedure for detecting the presence of viable bacteria in blood.

CAMPYLOBACTER
A curved Gram negative non-sporing bacterium.

CASE CONTROL STUDY
An epidemiological study in which the characteristics of persons with disease (e.g. their food histories) are compared with a matched control group of persons without the disease or infection.

CHOLERA-LIKE TOXIN
An entero-toxin which has similar effects to that produced by strains of *Vibrio cholerae* responsible for the symptoms of cholera, which include a profuse dehydrating diarrhoea.

COLONISATION
The phenomenon of a population of micro-organisms becoming established in a certain environment (especially in the intestinal tract of humans or animals) without necessarily giving rise to disease.

COMMENSAL
An organism which derives benefit from living in close physical association with another organism, the host, which derives neither benefit nor harm from its relationship with the commensal.

COMPETITIVE EXCLUSION
Growth of two different bacterial species in competition with each other which results in decline of one species and increase in the other.

CULTURE MEDIUM
A liquid or solid medium which is capable of supporting the growth of micro-organisms.

D-VALUE
The time required at a given temperature to reduce the number of viable cells or spores of a given micro-organism to 10% of the initial number, usually quoted in minutes.

ENTERITIS
Inflammation of the intestine.

ENTEROPATHOGEN
A pathogen that can cause disease in the intestines.

EPIDEMIOLOGY
The study of factors affecting health and disease in populations and the application of this study to the control and prevention of disease.

EPITHELIAL CELLS
Cells which form the layer (the epithelium) which lines the inner surface of the intestines.

EXTRA-INTESTINAL
Outside the intestines.

GENOTYPING
Distinguishing organisms by their content of genetic information, either in total or with respect to particular factors, regardless of whether the information in the genetic material is expressed in the characteristics of the organism.

GENUS
A sub-class of organisms which have natural affinities or similarities and can be sub-divided into the species of that genus.

GRAM NEGATIVE
A reaction of a staining procedure used as an initial step in the identification of bacteria.

GUILLAIN-BARRÉ SYNDROME
A disorder characterised by acute onset of weakness in the distal muscles of the legs which spreads upwards over the course of a few days to involve the trunk, arms and sometimes the cranial nerves.

IgA, IgG, IgM
Different types of immunoglobulin found in body fluids.

IMMUNE STATUS
State of immunity or resistance to infection of an individual to a particular pathogen.

IMMUNOCOMPROMISED
An individual who is unable to mount a normal immune response.

IMMUNOGLOBULINS
A class of proteins which are antibodies and found in body fluids.

INCUBATION PERIOD
The time interval between the initial entry of a pathogen into a host, and the appearance of the first symptoms of disease.

INDEX CASE
The first case in an outbreak of infectious disease.

INFECTIOUS DOSE
The amount of infectious material, e.g. number of bacteria, necessary to produce an infection.

INTESTINAL FLORA
Commensal organisms living in the intestine.

ISOLATE
Bacterial growth originating from a particular sample.

MASTITIS
Inflammation of the mammary gland.

MENINGES
The membranes surrounding the brain and the spinal cord.

MICROAEROBIC
Refers to a gaseous environment in which oxygen is present but is at a concentration (partial pressure) significantly lower than in air. A microaerobic organism prefers or can only survive in such an environment.

MOTILE STRAINS
Bacterial strains which can move independently.

NEONATAL SEPSIS
The condition in which a new born baby has symptoms associated with microbial infection of tissues.

NEONATE
A new born baby, up to four weeks of age.

NEUROLOGICAL COMPLICATIONS
Symptoms which occur in the nervous system as a complication of a disease which primarily affects another part of the body.

PASTEURISATION
A form of heat treatment that kills vegetative pathogens and spoilage bacteria in milk and other foods.

PATHOGEN
Any biological agent that can cause disease.

PATHOGENICITY
Ability to behave as a pathogen.

pH
An index used as a measure of acidity or alkalinity.

REACTIVE ARTHRITIS
A non-infective arthritis which may be secondary to an episode of infection elsewhere in the body.

SECRETORY IgA

A form of IgA which is resistant to enzymic breakdown and is found on mucosal surfaces e.g. the lumen of the gastrointestinal tract.

SELECTIVE MEDIA
Types of culture media which use selective agents such as antibiotics to inhibit some types of bacteria to allow the growth of others.

SENTINEL PRACTICE SCHEME
The Royal College of General Practitioners Research Unit's reporting scheme for a wide range of clinical diagnoses including infectious gastrointestinal disease.

SEROTYPING
A method of distinguishing varieties of bacteria by defining their antigenic properties.

SERUM ANTIBODIES
Antibodies found in the fluid fraction of coagulated blood.

SPECIES
A classification of organisms within a genus which have similarities and can be further sub-divided into sub-species.

SPECIFIC IMMUNE RESPONSE
Any form of immune response which is specific to a given antigen.

SPORADIC CASE
A single case of disease apparently unrelated to other cases.

SPORING/NON-SPORING
Refers to the potential/lack of potential of an organism to produce an environmentally resistant form called a spore.

STRAIN
A population of organisms within a species or sub-species distinguished by sub-typing.

SUB-SPECIES
A classification of organisms within a species which have similarities.

SUB-TYPING
Any method used to distinguish between species or sub-species.

SUSCEPTIBLE INDIVIDUAL
An individual who has no pre-existing immunity or resistance to infection who is therefore liable to become infected.

THERMOPHILIC GROUP
Refers to those campylobacters which grow well at 42°C and 37°C, but not at 25°C.

TYPING
Any method used to distinguish between closely related micro-organisms.

UNTYPABLE
Refers to an organism which is not able to be differentiated from closely related organisms by a particular typing method.

VEGETATIVE CELLS
Cells in which nutrition and growth predominate.

VERTICAL TRANSMISSION
The transmission of a disease or parasite from a parent to its offspring via the egg, via the placenta, or by genetic inheritance.

VIABLE
Refers in microbiology to an organism capable of reproducing under appropriate conditions.

CULTURABLE/NON-CULTURABLE
Refers to an organism which can/cannot currently be grown on a culture medium.

VIRULENCE FACTOR
A factor affecting the ability of an organism to cause disease. Virulence is defined broadly in terms of the severity of the symptoms in the host. Thus a highly virulent strain may cause severe symptoms in a susceptible individual, while a less virulent strain would produce relatively less severe symptoms in the same individual.

WATER ACTIVITY A_w
A measure of the available water in a substance.

APPENDIX 1

THE GENUS *CAMPYLOBACTER*

Genus definition

A 1.1 Members of the genus *Campylobacter* are slender, spirally curved rods between 0.2 μm to 0.5 μm wide and 1.5 μm to 5 μm long. Campylobacters are non-sporing Gram negative Eubacteria with a single polar flagellum whose action results in a corkscrew-like motility. Some species can be differentiated on the basis of the characteristic wave-length of their helical cells, but under certain cultural conditions and in old cultures, cells may become coccoidal.[171, 172] This morphological change may be associated with the transition to a viable, non-culturable form.[8]

A 1.2 All campylobacters have a guanine and cytosine content of 29-38% and are oxidase positive, nitrate positive, the one exception being *C.jejuni* subspecies *doylei*, which is oxidase positive but nitrate negative.[173] Campylobacters neither ferment nor oxidise carbohydrates but obtain energy from oxidation of amino acids or tricarboxylic acid intermediates. This means that many conventional microbiological identity tests are not applicable. The genus is commonly described as microaerobic because oxygen at normal atmospheric pressure is toxic to growth.[12] This is due to the production of hydrogen peroxide and superoxide anions which are toxic forms of oxygen for most campylobacters, but particularly to the catalase negative or weakly positive species. Adaption of *C. jejuni* to aerobic metabolism has been described.[174]

A 1.3 The MAFF Predictive Modelling Programme can now provide information using a survival model for *C.jejuni* based on the effect of temperature, pH and sodium chloride concentration. The range of conditions within the matrix are 5-25°C, pH4.5-7.0, and 0.5-4.5 sodium chloride (% w/v). The following observations were made from this data:

- survival time increases as temperature declines;
- survival time increases as pH increases; and
- survival time increases as the sodium chloride content decreases (Personal communication from Dr J P Sutherland, AFRC Institute of Food Research, Reading Laboratory)

Figures A1.1, A1.2, A1.3 are 3-D plots for *C.jejuni* showing how the D-value changes when pH and temperature are varied in different salt concentrations.

Figure A1.1

C. jejuni : Variation in D-value (days) with temperature and pH value (0.5% NaCl)

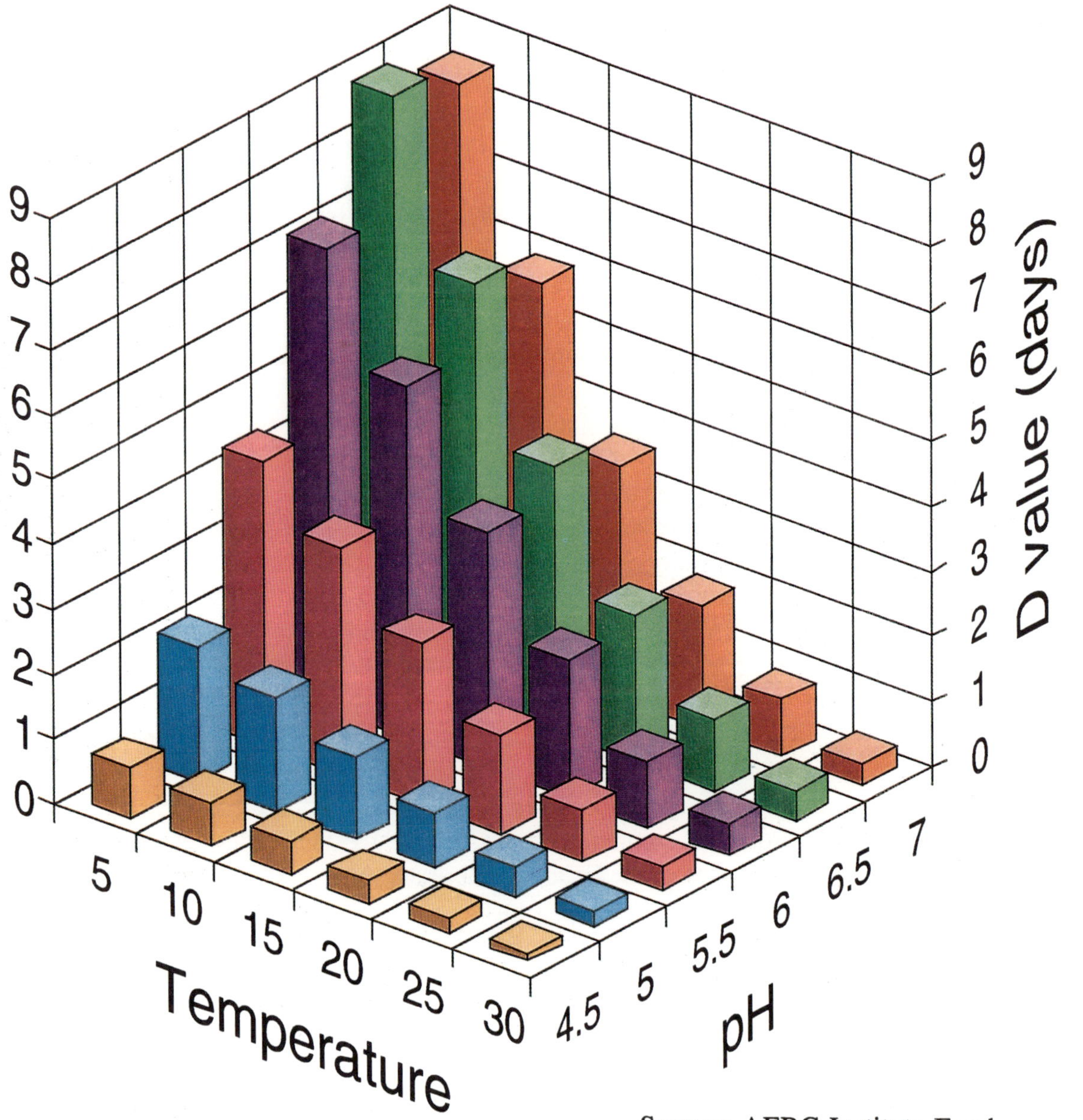

Source: AFRC Institute Food Research, Reading Laboratory
Prepared by Dr J P Sutherland

Figure A1.2

C. jejuni : Variation in D-value (days) with temperature and pH value (2.5% NaCl)

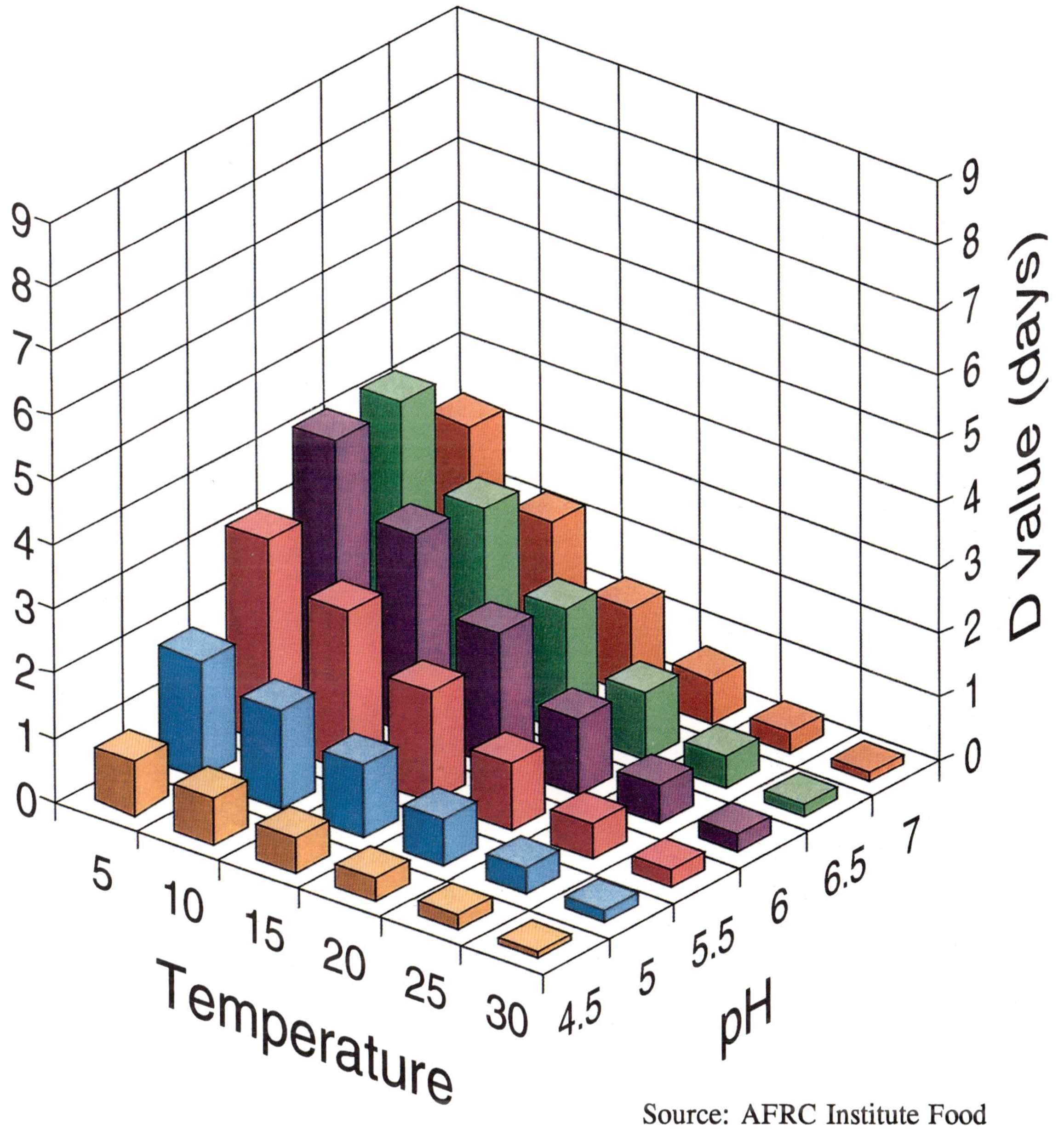

Source: AFRC Institute Food Research, Reading Laboratory
Prepared by Dr J P Sutherland

Figure A1.3

C. jejuni : Variation in D-value (days) with temperature and pH value (4.5% NaCl)

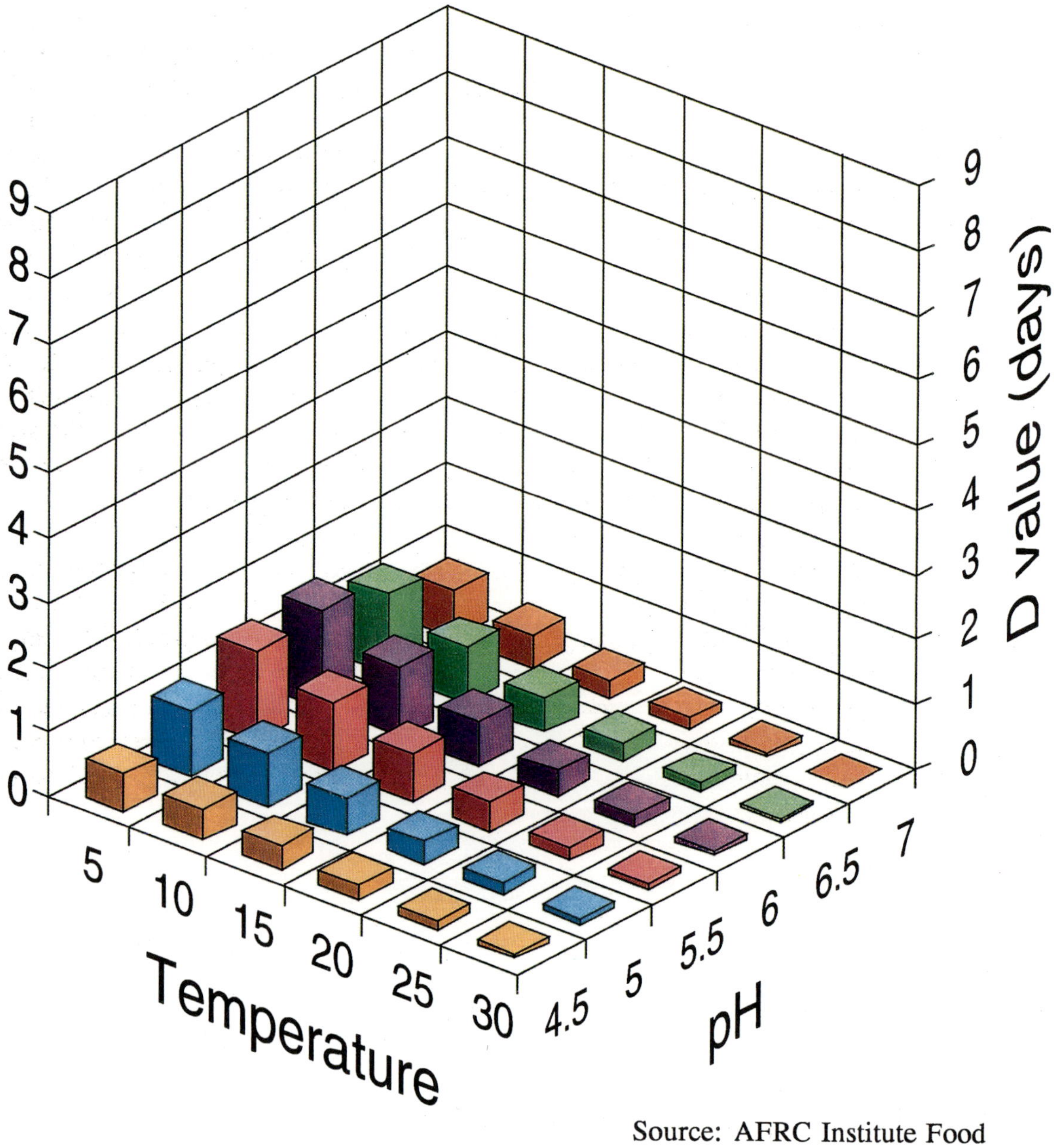

Source: AFRC Institute Food Research, Reading Laboratory
Prepared by Dr J P Sutherland

Species of importance in human disease

A 1.4 Differential properties of the established species and sub-species of the genus associated with human disease are shown in table A1.1. Recently three additional biochemical tests have been suggested.[175]

A 1.5 The natural habitat of all campylobacters is in the reproductive organs, intestinal tract and oral cavity of man and animals. Members of the genus associated with human disease differ with respect to their growth temperatures. *C.jejuni, C.coli, C.lari,* and *C.upsaliensis* are often referred to as the thermophilic group as they grow well at 42°C but not at 25°C. In contrast, *C.fetus* has an optimum growth temperature of 25°C, and *C.hyointestinalis* is able to grow at 25°C and 42°C. *C.lari* is a nalidixic acid resistant thermophilic campylobacter commonly isolated from seagulls and some other birds, but is an infrequent human pathogen.[23] *C.hyointestinalis* is a species found in the intestines of pigs and other animals which rarely causes disease in humans except in the immunocompromised.[24] *C.upsaliensis*, a catalase negative or weakly positive species, is associated with human enteritis, and is possibly as common in the UK as *C.coli* but is less commonly isolated by clinical laboratories.[25, 177, 178, 179] *C. butzleri* is a newly described aerotolerant species which is able to grow at 15°C, does not react in *C.jejuni/coli* serotyping schemes, and is distinctive by DNA hybridisation tests. The sparse epidemiological information to date suggests that *C.butzleri* may be associated with both sporadic cases and outbreaks of campylobacteriosis.[26, 27, 180] Recently *C.butzleri* has been moved to the genus *Arcobacter*,[181] and a new family, *Campylobacteraceae*, comprising both *Campylobacter* and *Arcobacter* species, has been proposed.[182]

A 1.6 In 1980 the most commonly isolated thermophilic species accounting for >90% of foodborne infections in the UK was *C.jejuni*, with *C.coli* causing only 2%,[183] although many clinical laboratories do not routinely differentiate between the two. In the period 1989-1990 only 17% of campylobacters reported to CDSC by laboratories in England and Wales were speciated. Of these 89.5% were *C.jejuni*, 10.3% were *C.coli* and 0.2% were other named agents. The currently available methods for sub-species typing are discussed in Appendix 2.

Table A1.1

BIOCHEMICAL AND GROWTH CHARACTERISTICS OF *CAMPYLOBACTER* SPECIES ASSOCIATED WITH HUMAN ENTERITIS

	*C. fetus**	*C. jejuni*	*C. coli*	*C. lari*	*C. upsaliensis*	*C. hyointestinalis*
Growth at:						
25°C	+	–	–	–	–	w
42 or 43°C	–	+	+	+	d	+
Anaerobic growth with trimethylamine-N-oxide	–	–	–	+	–	+
Catalase	+	+/w	+	+	-/w	+
Nitrate reduction	+	d	+	+	+	+
Hippurate hydrolysis	–	+	–	–	–	–
Indoxyl acetate hydrolysis	–	+	+	–	+	–
Growth in 1% glycine	+	+	+	+	–	+
H_2S production in TSI	–	d	d	+^	–	+$
Sensitivity to:						
Nalidixic acid	R	S	S	R	S	R
Cephalothin	S	S$	R	R	S	S
Selenite reduction	d	d	+	d	+	+
Growth on potato starch	N	N	N	N	+	N
G + C content (mol%)	33-36	29-32	31-33	31-33	35-36	35-36

Test results: + = positive reaction; - = negative reaction; w = weak reaction; d = some positive, others negative; N = not known; S = sensitive; R = resistant.

TSI: triple sugar iron medium.
$: occasional strain giving atypical result
^ : at 3 days, negative result
* : included as type-species for the genus

Adapted from Stanley et al[176]

Isolation

A 1.7 The species we now know as *C.jejuni/coli* were first isolated on solid medium using blood from humans with diarrhoea as the inoculum.[184] In 1972, the organisms were first isolated from stools by a procedure which involved passing diluted stools through a 0.65 μm Millipore filter, and then plating out on blood-thioglycollate medium containing bacitracin, polymyxin, novobiocin, and actidione.[6] In 1977, Skirrow[18] described a selective isolation technique which involved direct culture of faeces onto blood agar containing vancomycin, polymyxin and trimethoprim, and incubation at 43°C in an atmosphere of 5% oxygen, 10% carbon dioxide, and 85% hydrogen. This simplified isolation technique has enabled routine diagnostic laboratories to isolate *C.jejuni/coli* from stool specimens. Most clinical laboratories still use the microaerobic and thermophilic properties of *C.jejuni/coli* as selective characteristics in primary isolation, but many have now changed to using a blood-free medium (see Table A1.2). Many laboratories use 48 hours incubation to recover *Campylobacter* species, but there are reports that 72 hours or longer improves recovery.[185]

A 1.8 A comparative study of six different media showed that the highest isolation rates were obtained when two media were combined.[186] The composition of commonly used selective media for the isolation of *C.jejuni/coli* is shown in Table A1.2. Isolation of *C.jejuni/coli* from stools of patients with diarrhoea is relatively easy compared to their isolation from stool samples of convalescents, or from patients post antibiotic treatment, or from food. In such cases the use of an enrichment medium has improved isolation rates.[20, 187, 188] The growth of *C.coli* and some of the more recently described species, e.g. *C.upsaliensis*, has been found to be inhibited by the presence of antibiotics in selective medium.[22, 193, 194] A non-inhibitory filtration method of isolation may be needed so that a clearer picture of the role of different *Campylobacter* species in human enteritis can be determined.

A 1.9 In 1986 Rollins and Colwell[8] reported the existence of viable non-culturable forms of *C.jejuni* in a natural aquatic environment. They found that stream water held at 4°C sustained significant numbers of campylobacters for more than 4 months. During transition from the culturable to non-culturable state, a morphological change to coccoidal forms took place. Reports of recovery of viable, non-culturable *C.jejuni* by passage through the mouse and rat intestinal tract have been published.[9, 195] Poultry can also be infected by the organism in this form.[10, 11] In contrast, there are reports indicating a lack of colonisation by viable, non-culturable *C.jejuni* in one day old chicks, mice and rabbits.[11, 96]

Table A1.2

COMPOSITION OF COMMONLY USED SELECTIVE MEDIA USED FOR ISOLATION OF *C.JEJUNI/COLI*

Medium	Base	Selective agent	Supplements	Reference
Skirrows	Blood agar base No:2 (Oxoid CM 271)	Vancomycin 10 mg/l Polymyxin B 2500iu/l Trimethoprim 5 mg/l	Lysed horse blood 50ml/l	Skirrow[18]
Preston	Nutrient broth No:2 (Oxoid CM 67) New Zealand agar	Rifampicin 10mg/l Polymyxin B 5000iu/l Trimethoprim 10mg/l Cyclohexamide 100mg/l	Lysed horse blood 50ml/l	Bolton and Robertson[189]
Modified CCDA	Blood free selective agar base (Oxoid CM 739)	Cefoperazone 32mg/l		Hutchinson and Bolton[190]
Butzler Medium Virion	Columbia agar (Oxoid CM 331)	Cefoperazone 15mg/l Rifampicin 10mg/l Colistin 10000iu/l Amphotericin B 2mg/l	Sheep blood 50-70ml/l	Goosens et al[191]

Adapted from Griffiths and Park[192]

A 1.10 Although *Campylobacter* species of importance in human enteritis can be differentiated by biochemical and growth characteristics, newer methods involve the use of DNA probes and the polymerase chain reaction (PCR). Taylor and Hiratsuka[197] reported that a non-radioactive DNA probe, which contained a base pair (bp) fragment specific for the major outer membrane protein of *C.jejuni*, demonstrated a sensitivity of 93% with culture positive stool specimens. In another study of over 400 organisms, a DNA probe was used to identify *C.jejuni*, with 100% accuracy.[198] Using primers specific for the flagellin genes or 16S ribosomal RNA (rRNA) genes, PCR has been used to detect *C.jejuni* and *C.coli* directly in stool samples and chicken samples.[199,200,201]

APPENDIX 2

SUB-SPECIES TYPING OF *C.JEJUNI/COLI*

Biotyping

A 2.1 Skirrow and Benjamin[202] were the first to report an extensive biotyping survey of over a thousand campylobacters isolated from a variety of animal and environmental sources. Their most useful differential tests were growth at 25, 30.5, 37, 43 and 45.5°C, and tolerance tests to nalidixic acid, 2,3,5,triphenyltetrazolium chloride, 1.5% sodium chloride and metronidazole. From the results of these tests they were able to distinguish *C.fetus* subspecies *fetus, C.jejuni, C.coli*, and nalidixic acid resistant thermophilic campylobacters (NARTC). These same workers soon afterwards published a short communication on the differentiation of enteropathogenic campylobacters using only five tests, growth at 25°C and 43°C, growth in presence of nalidixic acid, hippurate hydrolysis and the production of hydrogen sulphide in iron medium.[203] Using these tests it was possible to distinguish *C.fetus*, *C.jejuni* biotype 1, *C.jejuni* biotype 2, *C.coli*, and NARTC. The biotyping of *C.jejuni* and *C.coli* was subsequently extended by the use of a DNA hydrolysis test; this distinguished four biotypes of *C.jejuni*, I, II, III, and IV, and two biotypes of *C.coli*, I, and II.[204] An improved biotyping scheme for *C.jejuni* and *C.coli* was also reported in 1984 by Roop et al,[205] which used only four tests, namely alkaline phosphatase activity, DNase activity, hippurate hydrolysis, and the ability to grow in a minimal medium. Roop and his colleagues assigned 32 strains to one of 4 biovars of *C.jejuni*, or to one of 4 biovars of *C.coli*.

A 2.2 Probably the most useful biotyping scheme to date is that developed by Bolton et al.[28] This scheme consists of testing the susceptibility of campylobacters to various antibiotics, dyes and chemicals in 10 resistotyping tests, together with their ability to grow at 25°C and to hydrolyse hippurate. This Preston biotyping scheme is able to speciate and biotype campylobacters, the end result being the designation of a species name and a four figure biotype code to an isolate.

Bacteriophage typing

A 2.3 Other sub-typing schemes have been devised using the lytic patterns produced by bacteriophages. Grajewski et al[31] have produced such a scheme by analysing the lytic patterns of 47 'phages on 150 randomly selected *C.jejuni/coli* human isolates. Numerical taxonomic techniques were used to assist in the selection of 14 'phages for a final typing set which was initially tested on 255 human *C.jejuni/coli* isolates; 95% percent of isolates were found to be typable and 46% fell within the four most common 'phage patterns. A recent extended 'phage typing scheme has been described which was used to demonstrate that the common human 'phage types of *C.jejuni* and *C.coli* were also found in cattle and poultry.[206]

Serotyping

A 2.4 Penner and Hennessy[30] devised the first serotyping scheme for *C.jejuni* in 1980. Using patterns of reactivity of red blood cells (RBCs) sensitised with extracted *Campylobacter* antigen with rabbit antisera raised against reference strains, it is possible to type a *C.jejuni/coli* isolate using a passive haemagglutination assay. The antigens involved in the passive haemagglutination typing scheme are now known to be lipopolysaccharide (LPS), referred to as 'O' antigens in other Gram negative bacteria. This serotyping method is referred to as the Penner or heat-stable (HS) scheme and can identify 60 serotypes (42 *C.jejuni* and 18 *C.coli*). The most common Penner serotypes isolated over a seven year study in Manchester were 4,2,1,6,11,9, in descending order of frequency, with remaining serotypes occurring at a frequency of less than 3%.[207] Commercial antisera for serotyping heat-labile antigens of *C. jejuni* and *C. coli* has recently been evaluated.[208] An alternative approach to serotyping *C.jejuni* was reported by Lior et al.[29] The Lior scheme was based on a rapid slide agglutination technique using live bacteria for the detection of heat-labile (HL) antigens, and can identify 108 serotypes.[35] This typing scheme has not been so extensively used in the UK, but in typing over 3,000 human isolates in Canada, Lior and Woodward[209] found *C.jejuni* Lior serotype 4 to be the most common, followed by Lior serotypes 7, 1, 2, 36, 5, 9, 17, 8, 6, 11, 18, 15, 16 and 28.

A 2.5 It has also been possible to define other serotypes of *C.jejuni* and *C.coli* by direct immunofluorescence and by using bactericidal antibodies.[210, 211] Although several methods of serotyping *C.jejuni/coli* have been described, the Lior and Penner schemes continue to be those most widely used. Comparison of these two methods on over 1,000 isolates from human, animal and environmental sources showed that 96.1% were typable by the Penner and 92.1% by the Lior method.[212] In the two systems most strains assigned to a single serotype by the Penner method corresponded to one serotype by the Lior method and vice versa. However, some strains typed as multiple serotypes by one method and a single serotype in the other. As there are now over 100 Lior serotypes and over 70 Penner serotypes, a combination of the two schemes is needed for epidemiological purposes.

Genotyping methods

Chromosomal DNA fingerprinting

A 2.6 Chromosomal DNA fingerprinting refers to restriction fragment length polymorphism (RFLP) analysis of the bacterial genome. The application of this technique to distinguish different bacterial strains has the advantage of being very sensitive to minor genomic variations. This can be an asset in epidemiological studies to trace source isolates.

A 2.7 One of the disadvantages of chromosomal DNA fingerprinting is the generation of complex multiband fragment patterns. This means that defining visual similarities becomes rather subjective. Similar problems are encountered in the differentiation of *Campylobacter* species by comparison of whole cell protein banding patterns in SDS-PAGE.[213, 214] However, computer-assisted numerical analysis has made species and sub-species identifications possible.[215]

A 2.8 Similarly, numerical analysis has been useful in the interpretation of DNA fingerprints. Photographs of restriction patterns can be scanned with a laser densitometer, peak areas stored on disc, and similarities between all possible pairs of patterns computed. Using this approach it has been possible to produce dendrograms of cluster analysis showing the degree of similarity between *Campylobacter* strains and species.[216]

Ribotyping

A 2.9 A second type of DNA fingerprint that can be used in *Campylobacter* strain identification is based on the detection of RFLP in rRNA genes. The advantage of such RFLP patterns is that they are less complex than those obtained using total genomic DNA, easier to interpret, and also avoid any complications due to the presence of plasmid DNA.

A 2.10 A comparison of total DNA digests using *Hae* III and *Hind* III and their 16S and 23S rRNA gene patterns (ribopatterns) was carried out for a number of *C.jejuni* isolates obtained from three different outbreaks of enteritis.[217] The total DNA digest patterns yielded in excess of 30 bands by agarose gel electrophoresis; visual inspection enabled 4 distinct patterns to be discerned but the patterns were too complex for further analysis. The endonuclease digested genomic DNA was analysed by Southern blotting using a probe specific for 16S and 23S rRNA from *E.coli*. This gave much less complex results than total DNA digests, enabling comparisons to be made. Clear differences were found between the three sets of outbreak-linked strains of *C.jejuni*. In addition the results correlated with those obtained by serotyping and 'phage typing. Further work using both *C. jejuni* and *C. coli*, which included 53 different Penner serotypes, supports the discriminating power of *Hae* III ribotyping.[218]

A 2.11 Moureau et al[219] purposely chose a short synthetic DNA probe, coding for the end of 16S rRNA, conserved between *E.coli* and *C.jejuni* with the assumption that it would hybridise with the DNA of all *Campylobacter* species. After total DNA digestion with *Xho*I and *Bgl*II and Southern blot analysis using this probe, they found that each *Campylobacter* species including *C.jejuni, C.coli, C.fetus*, *C.lari* and *C.upsaliensis* displayed RFLP. The major advantage of DNA grouping techniques is that it is always possible to obtain a result, whereas some strains may prove indistinguishable or untypable by serotyping or 'phage typing.

Pulsed field gel electrophoresis (PFGE)

A 2.12 By using restriction enzymes that cut at few sites, a small number of large

DNA fragments can be produced from genomic DNA. PFGE is used for the electrophoretic separation of such large molecules of DNA in a gel matrix by alternating pulses of electric current in different directions. Using this method, genome maps of *C.jejuni* and *C.coli* have been produced.[220] Yan et al[221] used PFGE analysis of *Sma*I restricted genomic DNA to differentiate between *Campylobacter* species and found intraspecies differences, which suggests that the technique may be useful as an epidemiological tool.

Multi-locus enzyme electrophoresis (MEE)

A 2.13 In this technique, allelic variation in a structural gene is deduced by assessing variation in the net electrostatic charge of its polypeptide product. The proteins encoded by different structural genes are represented by enzymes of cellular metabolism such as isocitrate dehydrogenase, fumarase, alkaline phosphatase and catalase. Aeschbacher and Piffaretti[222] used this technique to analyse 125 *C.jejuni/coli* strains of both human and animal origin. 64 distinct electrophoretic types were distinguishable, 50 for *C.jejuni* and 21 for *C.coli.* For most of the enzymes studied allelic variation and therefore genetic diversity, was high. Results comparing the electrophoretic types of *C.jejuni* and *C.coli* confirmed that these represent two distinct species, as there was limited sharing of common alleles. Aeschbacher and Piffaretti also showed that their human *C.jejuni/coli* isolates were genetically indistinct from those of animal origin, supporting the hypothesis that *Campylobacter* enteritis is a zoonosis.

Use of polymerase chain reaction (PCR) in sub-species typing

A 2.14 Primer directed enzymic amplification of target DNA is an *in vitro* method of synthesing millions of copies of a nucleotide sequence in a few hours.[223] When a gene sequence has been determined it is possible to design primers to amplify a specified gene. After PCR amplification the resulting DNA can then be analysed for RFLPs.[200] Alternatively, arbitrary primers may be used which anneal to various unspecified sites along the DNA. This technique is referred to as randomly amplified polymorphic DNA (RAPD) analysis. Restriction enzymes are not needed in this technique as many short strands of DNA are produced which are then analysed by electrophoresis. Mazurier et al[224] used RAPD to analyse a range of *C.jejuni* serotypes and found that, depending on the primer used, different levels of discrimination between strains was possible. Correlation was demonstrated between RAPD analysis and serotyping, as was discrimination of strains within given Penner and Lior serotypes.

A 2.15 The most promising genotyping methods at the moment, in terms of providing a useful epidemiological strain differentiation combined with less complex methodology and analysis of results, are ribotyping and RAPD.

APPENDIX 3

PATHOGENICITY DETERMINANTS

Motility and mucus colonisation

A 3.1 The action of the *C.jejuni/coli* flagella results in a corkscrew-like motility, which facilitates penetration of the intestinal mucus and allows colonisation of the intestinal tract.[15] Some *C.jejuni/coli* strains undergo a bidirectional transition between flagellated and unflagellated forms,[59, 225] but the flagellate phenotype appears to be favoured by *in vivo* passage in both animals and humans.[59, 60] In a study with human volunteers, despite ingestion of equal numbers of both motile and non-motile variants of one strain, only motile organisms were isolated from stool samples.[39] Motility was cited by Morooka et al[58] as a virulence factor because wild-type motile strains of *C.jejuni* were found to colonise successfully in an animal model, but non-motile strains were cleared from the intestinal tract.

A 3.2 Although the mucus gel is considered to be a protective barrier for the intestinal epithelium, pathogens are able to overcome this barrier and gain access to the epithelium, often aided by tissue derived chemotactic factors.[57] Using an animal model Lee et al[226] reported that human *C.jejuni* isolates colonised mucus deep within the intestinal crypts of the caecum. Hence close proximity of the organisms to the epithelium could account for harmful effects on tissue via the production of enterotoxins or other irritants.

Adherence and invasion

A 3.3 The clinical presentation of blood-stained stools demonstrates that adherence and invasion of the intestinal epithelium may occur. To investigate this many *in vitro* adhesion and invasion assays for *C.jejuni* have been developed using a variety of different cell lines and detection systems.[62, 227]

A 3.4 Adherence has been found to be partially inhibited in the presence of various substances, a finding that suggests multiple adhesins are involved. Results from adherence inhibition assays and experiments using trypsin, periodate and heat-treated (60°C) bacteria suggested that the bacterial components involved in attachment and invasion may be protein, glycoprotein, or LPS in nature.[228, 229, 230] Indeed two types of bacterial components may be necessary, one for attachment and one for invasion.

A 3.5 Several studies have indicated a role for flagella in attachment to epithelial cells. Shearing of bacterial cells to remove flagella was shown to reduce adherence, whereas adherence was increased when flagella were immobilised artificially or non-motile flagellate strains used.[60, 229] More recent experiments with genetically modified mutants have clearly demonstrated the role of the flagellum during attachment and invasion.[61]

A 3.6 Invasion and intracellular survival studies of *C.jejuni* suggest that campylobacters attach to the cell membrane, are internalised by a phagocytic-like mechanism, and are digested after phagosome-lysosome fusion.[231] There is also evidence that *C. jejuni* is able to survive inside and have a toxic effect on epithelial cells.[232]

A 3.7 Evidence from some *in vitro* studies appears to indicate that clinical isolates of *C.jejuni* are more invasive than non-clinical isolates.[230, 233] However, more recent studies have shown that repeated *in vitro* sub-culturing may affect both antigen expression and invasive properties of strains.[234, 235]

Toxins

A 3.8 *Campylobacter* enteritis may take the form of a secretory diarrhoea which is particularly prevalent in young children in developing countries,[42] and is thought to be associated with the production of toxins by the infecting strain. Ruiz-Palacios[63] and colleagues first reported the detection of toxin production in 24 out of 32 *C.jejuni* strains from Mexican patients with diarrhoea. These initial observations were quickly supported by research workers in other parts of the world who also identified toxin producing *C.jejuni/coli* strains.[65, 236]

A 3.9 Supernatant from a prototype strain induced fluid secretion in a rat ileal loop, and elongation and an increased level of cAMP in Chinese hamster ovary cells (CHO). This toxin activity has since been found to be blocked by treatment of bacterial cells with GM1 ganglioside, or by anti-serum to the labile toxin of *E.coli* or cholera toxin and is destroyed by heat or high or low pH.[65, 66]

A 3.10 In addition to the cholera-like enterotoxin, other toxins have been reported including one which has a cytotoxic effect on Vero and HeLa cells, and a distinct cytolethal distending toxin.[237, 238] There are, however, contradictory reports in the scientific literature as to whether toxin production differs in strains from different sources.[64, 67]

A 3.11 Correlation of the pathogenic properties of the infecting *C.jejuni/coli* isolate and the gastrointestinal manifestations in the infected host has been attempted. In one study, isolates from individuals with inflammatory type diarrhoea all produced cytotoxins but not enterotoxins, whereas those from individuals with secretory type diarrhoea produced enterotoxins, and one produced a cytotoxin.[237] In another study correlation of toxin production with either secretory or inflammatory diarrhoea was not evident.[239]

A 3.12 Immunological evidence supports results from the latter study, as only one seroconversion was found when using an ELISA designed to detect *C.jejuni* specific IgG against cholera-like enterotoxin in 64 Mexican patients with inflammatory diarrhoea.[240]

A 3.13 In other pathogens, for example *E.coli*, plasmids specify a variety of virulence factors including enterotoxin production. As the presence of plasmids does not always correlate with enterotoxin production in *C.jejuni*, toxin production at least in some strains must be chromosomally encoded.[241]

A 3.14 However, some recent studies have failed to detect enterotoxin production or the presence of enterotoxin genes in a number of *Campylobacter* strains isolated from patients with inflammatory and secretory diarrhoea from both developing and developed countries.[44, 68]

A 3.15 It is possible that titres of *C.jejuni* toxins may be lower than for those produced by other enteropathogens, and more sensitive tests may be needed for their detection. The pathogenic significance of *C.jejuni* toxins remains to be proven.

APPENDIX 4

HOST ANTIBODY RESPONSE TO *C.JEJUNI/COLI* INFECTION

A 4.1 Several techniques have been used to study the serological and mucosal antibody response to *C.jejuni/coli* infection.[242, 243, 244] Serum antibody response was measured in an ELISA by Blaser and Duncan[69] using acid extracted antigens from three different Penner type strains. Sera were tested for *C.jejuni* specific IgG, IgM, and IgA from healthy controls, patients with *Campylobacter* enteritis and from chronic raw milk drinkers. Rising titres for all three isotypes were detected in patients with enteritis, and to a lesser extent in healthy controls exposed to the same source vehicle. IgG and IgM levels were generally found to persist longer than IgA levels. Increased IgG levels were found in chronic raw milk drinkers. A high prevalence of antibody in individuals who habitually consume raw milk and who presumably have multiple exposures to *C.jejuni* seems to lead to immunity, as these individuals suffer little illness.[70]

A 4.2 The detection of immunogenic components of *C.jejuni* during human infection was studied by comparing antibody activity in acute and convalescent sera from patients with *Campylobacter* enteritis. Detection of antibodies to LPS and the major outer membrane protein was variable, but a 66 kDa protein, identified as flagellin, reacted strongly with all sera tested.[245] When detecting *C.jejuni* flagellin serum responses in patients with campylobacteriosis, IgA and IgM were found to be more useful markers of recent infection than IgG.[246]

A 4.3 Serological studies carried out by Black et al,[39] during and after experimental *C.jejuni* infection in humans, indicated a peak serum antibody response for specific IgA at day 11 and at day 21 for IgM and IgG after infection. Several of the volunteers were rechallenged one month later and diarrhoea developed in none, suggesting that either the serological responses themselves are protective or may reflect other protective mechanisms such as the development of secretory IgA in the gut. Perez-Perez et al,[44] while failing to detect any serologic response to a cholera-like toxin, confirmed IgA, IgG and IgM seroconversion to *C.jejuni* surface antigens in sporadic cases of diarrhoea in Denver and Bangkok.

A 4.4 In other studies patients with agammaglobulinaemia or AIDS were found to have more severe or recurrent *Campylobacter* species intestinal or systematic infections.[52, 247]

A 4.5 There is some evidence that *C.jejuni* specific serum antibody levels are higher in those individuals occupationally exposed to *Campylobacter*, such as slaughter house workers, than in controls.[70, 71] Such indicators of specific immunity may account for a higher rate of *Campylobacter* infection in holiday relief workers than in regular staff, which was recorded during an outbreak of *Campylobacter* enteritis among the staff of a poultry abattoir in Sweden.[248]

A 4.6 In the UK, asymptomatic excretion of *C.jejuni* is uncommon; in developing countries carriage rates as high as 17% have been reported.[18, 42] In a study designed to compare the immune status to *Campylobacter* of children in developed and developing countries, Blaser et al[72] determined *C.jejuni* specific IgA, IgG and IgM levels in healthy Bangladeshi and US children in age groups from <1 to 15 years. For each age group studied, specific serum antibody levels were significantly higher in the Bangladeshi children and were nearly absent in serum from the US children until they reached 5 years of age. Whereas little *C.jejuni* specific IgG antibody was acquired during childhood by the US children, in the Bangladeshi children both IgM and IgG levels peaked in the 2-4 year age group. In Bangladeshi children there was an age related increase in serum IgA which may indicate the development of intestinal secretory IgA and correlate with the acquisition of protective gut immunity.

A 4.7 Ruiz-Palacios et al[75] have correlated the presence of *C.jejuni* specific secretory IgA in human milk with the prevention of *Campylobacter* enteritis in 98 breast-fed Mexican babies. Attack rates of diarrhoea in children less than 6 months of age who were bottle-fed was 2 to 3 times greater than those children of the same age who were breast-fed. Secretory IgA titres were generally high in both colostrum and milk, and the frequency of campylobacteriosis in breast-fed babies correlated with an absence of specific secretory IgA in the milk they consumed. More recently, these same workers have shown that non-immunoglobulin factors such as receptor analogues in human milk may also play a role in the protection afforded to breast-fed babies.[249]

A 4.8 *In vitro* experiments support the theory that intestinal mucus and secretory IgA may both prevent attachment of *C.jejuni* to intestinal epithelium *in vivo*.[250] *Campylobacter* specific intestinal secretory IgA has been measured both in patients with and without campylobacteriosis. Patients with *Campylobacter* infection had a higher mean faecal secretory IgA level specific for *Campylobacter* flagellin than did healthy controls, thus providing evidence of a local mucosal immune response.[73] Studies in animals have found that protective immunity to *Campylobacter* is associated with the development of both intestinal IgA and serum anti-flagellin antibodies.[74, 251]

REFERENCES

1. Bryan F L. Foodborne infections and intoxications: contemporary problems and solutions. In: Foodborne infections and intoxications. Proceedings of the 3rd World Congress; 1992 June 16-19; Berlin. Berlin: Institute of Veterinary Medicine - Robert von Ostertag-Institute of the Federal Health Office (FAO/WHO Collaborating Centre for Research and Training in Food Hygiene and Zoonoses), 1992; 11-19.

2. Sockett P N, Pearson A D. Cost implications of human *Campylobacter* infections. In: *Campylobacter* VI. Editors: B Kaijser, E Falsen. Gothenberg: University of Gothenberg. 1988; 261-4.

3. The Microbiological Safety of Food: Part I. Report of the Committee on the Microbiological Safety of Food (Chairman: Sir Mark Richmond). London: HMSO, 1990.

4. The Microbiological Safety of Food: Part II. Report of the Committee on the Microbiological Safety of Food (Chairman: Sir Mark Richmond). London: HMSO, 1991.

5. McFadyean J, Stockman S. Report of the departmental committee appointed by Board of Agriculture and Fisheries to inquire into epizootic abortion. Appendix II and p22. HMSO, London, 1913.

6. Dekeyser P, Gossuin-Detrain M, Butzler J P, Sternon J. Acute enteritis due to related Vibrio: first positive stool culture. J. Infect. Dis. 1972; **125:** 390-392.

7. Levy A J. A gastroenteritis outbreak probably due to a bovine strain of vibrio. Yale J. Biol. Med. 1946; **18:** 243-251.

8. Rollins, D M, Colwell, R R. Viable but nonculturable stage of *C. jejuni* and its role in survival in the natural aquatic environment. Appl. Environ. Microbiol. 1986; **52:** 531-538.

9. Jones D M, Sutcliffe E M, Curry A. Recovery of viable but non-culturable *Campylobacter jejuni*. J. of Gen. Microb. 1991; **137:** 2477-82.

10. Pearson A D, Greenwood M, Healing T D, Rollins D, Shanamath, Donaldson J, Colwell R R. Colonisation of broiler chickens by waterborne *Campylobacter jejuni*. Appl. Environ. Microbiol. 1993; **59:** 987-996.

11. Medema G J, Schets F M, Van de Giessen A W, Havelaar A H. Lack of colonisation of 1 day old chicks by viable, non-culturable *Campylobacter jejuni*. J. Appl. Bacteriol. 1992; **72:** 512-6.

12. Hoffman P S, George H A, Krieg N R, Smibert R M. Studies of the microaerophilic nature of *Campylobacter fetus subsp jejuni*: II Role of exogenous superoxide anion and hydrogen peroxide. Can. J. Microbiol. 1979; **25:** 8-16.

13. Bolton F J, Coates D. A Study of the oxygen and carbon dioxide requirements of thermophilic campylobacters. J Clin Path. 1983; **36:** 829-834.

14. Fernie D S, Park R W A. The isolation and nature of *Campylobacters* /microaerophilic *Vibrios* from laboratory and wild rodents. J. Med. Microbiol. 1977; **10:** 325-329.

15. Stern N J, Kazmi S U. *Campylobacter jejuni*. In Foodborne Bacterial Pathogens. Editor; M P Doyle. Marcel Dekker, Inc. New York, 1989.

16. Hänninen M L. Effect of Nace in *Campylobacter jejuni/coli*. Acta. Vet. Scand. 1981; **22:** 578-588.

17. Cuk Z, Annan-Pak A, Jane M, Zajcsalter. Yogurt an unlikely source of *Campylobacter jejuni/coli*. J. Appl. Bacteriol. 1987; **63:** 201-205.

18. Skirrow M B. *Campylobacter* enteritis; a new disease. Br. Med. J. (Clin. Res. Ed.) 1977; **2:** 9-11.

19. Blaser M J, Hardesty H L, Wang W L L. Survival of *Campylobacter fetus* sub sp. *jejuni* in biological milieus. J. Clin. Microbiol. 1980; **11:** 309-13.

20. de Boer E, Humphrey T J. Comparison of methods for the isolation of thermophilic campylobacters from chicken products. Microb. Ecol. Health Dis. 1991; **4:** 543.

21. Scotter S L, Humphrey T J, Henley A. Methods for the detection of thermotolerant campylobacters in foods: results of an inter-laboratory study. J. Appl. Bacteriol 1993; **74:** 155-163.

22. Shanker S, Gordon S W, Fuller H, Gilbert G L. Sensitivity of a filtration method for detection of *Campylobacter/Helicobacter* species in faeces. Microb. Ecol. Health Dis. 1991; **4:** 547.

23. Benjamin J, Leaper S, Owen R J, Skirrow M B. Description of *Campylobacter* laridis, a new species comprising the nalidixic acid resistant thermophilic *Campylobacter group* (NARTC). Curr. Microbiol. 1983; **8:** 231-238.

24. Gebhart C J, Edmonds P, Ward G E, Kurtz H J, Brenner D J. *Campylobacter hyointestinalis sp nov*: a new species of *Campylobacter* found in the intestines of pigs and other animals. J. Clin. Microbiol. 1985; **21:** 715-720.

25. Bolton F J, Hutchinson D N, Parker G. Isolation of *Campylobacter*. What are we missing? J. Clin. Pathol. 1987; **40:** 702-703.

26. Kiehlbauch J A, Brenner D J, Nicholson M A, Baber C N, Patton C M, Steigerwalt A G, Wachsmuth I K. *Campylobacter butzleri* sp. nov. isolated from humans and animals with diarrhoeal illness. J. Clin. Microbiol. 1991; **29:** 376.

27. Kiehlbauch J A, Tauxe R J, Wachsmuth, I K. Clinical features of *Campylobacter butzleri* associated diarrheal illness. In: the IVth International Workshop on *Campylobacter, Helicobacter* and related organisms. Microbial Ecology in Health and Disease, **4:** special issue, 1991.

28. Bolton F J, Holt A V, Hutchinson. *Campylobacter* biotyping scheme of epidemiological value. J. Clin. Pathol. 1984; **37:** 677-681.

29. Lior H, Woodward D L, Edgar J A, Laroche L J, Gill P. Serotyping of *Campylobacter jejuni* by slide agglutination based on heat-labile antigenic factors. J. Clin. Microbiol. 1982; **15:** 761-768.

30. Penner J L, Hennessy J N. Passive haemagglutination technique for serotyping *Campylobacter fetus subsp jejuni* on the basis of soluble heat-stable antigens. J. Clin. Microbiol. 1980; **12:** 732-737.

31. Grajewski B A, Kusek J W, Gelfand H M. Development of a bacteriophage typing system for *Campylobacter jejuni/coli*. J. Clin. Microbiol. 1985; **22:** 13-18.

32. Jones D M, Fox A J, Eldridge J. Characterisation of the antigens involved in serotyping strains of *Campylobacter jejuni* by passive haemagglutination. Curr. Microbiol. 1984; **10:** 105-110.

33. Pearson A D, Healing T D, Sockett P, Jones D M (1989b). *Campylobacter*: the third agent in the food poisoning outbreak. In: *Salmonella* and *Listeria*. Editor: O Goldring. London EAG Scientific. 1989, 245-81.

34. Hutchinson D N, Bolton F J, Jones D M, Sutcliffe E M, Abbot J D. Application of three typing schemes (Penner, Lior, Preston) to strains of *Campylobacter* species isolated from three outbreaks. Epidemiol. Infect. 1987; **98:** 139-144.

35. Patton C M, Wachsmuth I K. Typing schemes: are current methods useful? In: *Campylobacter jejuni*: Current status and future trends. Editors: Nuchamkin I, Blaser M J, Tompkins L S. American Society for Microbiology. Washington DC, 1992.

36. Patton C M, Wachsmuth I K, Evins G M, Kiehlbauch J A, Plikaytis B O, Troup N, Tompkins L, Lior H. Evaluation of 10 methods to distinguish epidemic-associated *Campylobacter* strains. J Clin. Microbiol. 1991; **29:** 680-688.

37. Robinson D A. Infective dose of *Campylobacter jejuni* in milk. Br. Med. J. (Clin. Res. Ed) 1981a May 16; **282(6276):** 1584.

38. Robinson D A. *Campylobacter* infection. R. Soc. Health J. 1981b Aug; **101(4):** 138-140.

39. Black R E et al. Experimental *Campylobacter jejuni* infections in humans. J. Infect. Dis. 1988; **157:** 472-479.

40. Butzler J P, Glupczynski Y, Goossens H. *Campylobacter* and *Helicobacter* infections. Curr. Opin. Infect. Dis. 1992; **5:** 80-87.

41. Healing T D, Greenwood M H, Pearson A D. *Campylobacter* and enteritis. Rev. Med. Microbiol. 1992; **3:** 159-167.

42. Blaser M J, Taylor D N, Feldman R A. Epidemiology of *Campylobacter jejuni* infections. Epidemiologic Rev. 1983; **5:** 157-176.

43. Kapperud G, Lassen J, Ostroff S M, Aasen S. Clinical features of sporadic *Campylobacter* infections in Norway. Scand. J. Infect. Dis. 1992; **24:** 741-749.

44. Perez-Perez G I, Taylor D N, Echeverria P D, Blaser M J. Lack of evidence of enterotoxin involvement in pathogenesis of *Campylobacter* diarrhoea. In: *Campylobacter jejuni*; current status and future trends. 184-192, Editors: I. Nachamkin, M J Blaser, L S Tompkins. ASM Publication. Washington DC, 1992.

45. Anders B J, Lauer B A, Reller L B. Double-blind placebo controlled trial of erythromycin for the treatment of *Campylobacter* enteritis. Presented at 21st Interstate conference an anti-microbial agents and chemotherapy. Chicago, USA, 1981.

46. Pithie A D, Wood M J. Treatment of infectious diarrhoea with ciproflixacin. J. Antimicrob. Chemother. 26, Suppl. F. 1990; 47-53.

47. Bryan C S. Clinical implications of positive blood cultures. Clin. Microbiol. Rev. 1989; **2:** 329-353.

48. Skirrow M B. Foodborne illness. Lancet 1990; **336:** 921-923.

49. Kuroki S, Haruta T, Yoshioka M, Kobayashi Y, Nukina M, Nakanishi H. Guillain-Barre Syndrome associated with *Campylobacter* infection. Paediatr. Infect. Dis. 1991; **10:** 149-151.

50. Fujimoto S, Yuki N, Itoh T, Amako K. Specific serotype of *Campylobacter jejuni* associated with Guillain-Barre Syndrome. J. Infect. Dis. 1992; **165:** 183.

51. Blaser M J, Reller L B. *Campylobacter* enteritis. New. Engl. J. Med. 1981; **305:** 1444-1452.

52. Johnson R J, Nolan C, Wang S P, Shelton W R, Blaser M J. Persistance of *Campylobacter jejuni* in an immunocompromised patient. Ann. Intern. Med. 1984; **100:** 832-834.

53. Thomas K, Chan K N, Ribero C D. *Campylobacter jejuni* meningitis in a neonate. Br. Med. J. (Clin. Res. Ed.) 1980; **280:** 1301-1302.

54. Simor A E, Ferro S. *Campylobacter jejuni* infection occurring during pregnancy. Eur J. Clin. Microbiol. Infect. Dis. 1990 Feb; **9(2):** 142-144.

55. Denton K J. Role of *Campylobacter jejuni* as a placental pathogen. J. Clin. Pathol. 1992; **45:** 172-2.

56. Department of Health. While you are pregnant; Safe eating and how to avoid infection from food and animals. London; HMSO, 1992.

57. Freter R, Allweiss B, O'Brien P C M, Halstead S A, Macsai M S. Role of chemotaxis in the association of motile bacteria with intestinal mucosa: *in vitro* studies. Infect. Immun. 1981; **34:** 241-249.

58. Morooka T, Umeda A, Amako K. Motility as an intestinal colonisation factor for *Campylobacter jejuni*. J. Gen. Microbiol. 1985; **131:** 1973-1980.

59. Caldwell M B, Guerry P, Lee E C, Burans J P, Walker R I. Reversible expression of flagella in *Campylobacter jejuni*. Infect. Immun. 1985; **50:** 941-943.

60. Newell D G, McBride H, Dolby J. Investigations on the role of flagella in the colonisation of infant mice with *Campylobacter jejuni* and attachment of *Campylobacter jejuni* to human epithelial cell lines. J. Hyg. (Camb.) 1985; **95:** 217-227.

61. Wassenaar T M, Bleumink-Pluym N, Van der Zeijst B A M. Inactivation of *Campylobacter jejuni* flagellin genes by homologous recombination demonstrates that *fla*A but not *fla*B is required for invasion. EMBO J. 1991; **10:** 2055-2061.

62. Fauchere J L, Rosenau A, Veron M, Moyan E N, Richards S, Pfister A. Association with HeLa cells of *Campylobacter jejuni/coli* outer isolated from human faeces. Infect. Immun. 1986; **54:** 283-287.

63. Ruiz-Palacios G M, Torres J, Torres N L, Escamilla E, Ruiz-Palacios B R, Tamayo J. Cholera-like enterotoxin produced by *Campylobacter jejuni.* Lancet. 1983; 250-253.

64. Lindblom G-B, Kaijser, Sjogren E. Enterotoxin production and serogroups of *Campylobacter jejuni* and *Campylobacter coli* from patients with diarrhoea and from healthy laying hens. J. Clin. Microbiol. 1989; **27:** 1272-1276.

65. Johnson W M, Lior H. Cytotoxic and cytotonic factors produced by *Campylobacter jejuni, Campylobacter coli*, and *Campylobacter laridis*. J Clin. Microbiol. 1986; **24:** 275-281.

66. Klipstein E A, Engert R E. Properties of crude *Campylobacter jejuni* heat labile enterotoxin. Infect. Immun. 1984; **45:** 314-319.

67. McFarland B A, Neill S D. Profiles of toxin production by thermophilic *Campylobacter* of animal origin. Vet. Microbiol. 1992; **30:** 257-266.

68. Konkel M E, Lobet Y, Cieplak W. Examination of multiple isolates of *Campylobacter jejuni* for evidence of cholera toxin like activity. In *Campylobacter jejuni*: current status and future trends. Editors: Irving Nachamkin, Martin J Blaser, Lucy S Tompkins. Washington DC, American Society for Microbiology, 1992.

69. Blaser M J, Duncan D J. Human serum antibody response to *Campylobacter jejuni* infection as measured in an enzyme linked immunosorbent assay. Infect. Immun. 1984; **44:** 292-298.

70. Jones D M, Robinson D A, Eldridge J. Serological studies in two outbreaks of *Campylobacter jejuni* infection. J. Hyg. (Camb.) 1981; **87:** 163-170.

71. Mancinelli S, Palombi L, Riccardi F, Marazzi M C. Serological study of *Campylobacter jejuni* infection in slaughterhouse workers. J. Infect. Dis. 1987; **156:** 856.

72. Blaser M J, Black R E, Duncan D J, Amer J. *Campylobacter jejuni* specific serum antibodies are elevated in healthy Bangladeshi children. J Clin Microbiol 1985; **21:** 164-167.

73. Nachamkin I, Yang X H. Local immune responses to *Campylobacter* flagellin in acute *Campylobacter* gastrointestinal infection. J. Clin. Microbiol. 1992; **30:** 509-11.

74. Pavlovski O R, Rollins D M, Harberberger R L, Green A E, Habash L, Strocko S, Walker R J. Significance of flagella colonisation resistance of rabbits immunised with *Campylobacter* species. Infect. Immun. 1991; **59:** 2259-2264.

75. Ruiz-Palacios G M, Calva J J, Picking L K, Lopez V Y, Volkow P, Pezzarossi h, West M S (1990). Protection of breast-fed infants against *Campylobacter* diarrhoea by antibodies in human milk. J Paediatr. 1990; **116:** 707-713.

76. Kendall E J C, Tanner E I. *Campylobacter* enteritis in general practice, J. Hyg. (Camb.). 1982; **88:** 155-163.

77. Skirrow M J. A demographic survey of *Campylobacter*, *Salmonella* and *Shigella* infections in England. Epidemiol. Infect. 1987; **99:** 647-657.

78. Stern N J, Hernandez M P, Blankenship L et al. Prevalence and distribution of *Campylobacter jejuni* and *Campylobacter coli* in retail meats. J. Food Prot. 1985; **48:** 595-599.

79. Fricker C R, Park RWA. A two year study of the distribution of 'thermophilic' campylobacters in human, environmental and food samples from the Reading area with particular reference to toxin production and heat stable serotype. J. Appl. Bacteriol. 1989; **66:** 477-90.

80. Humphrey T J, Beckett P. *Campylobacter jejuni* in dairy cows and raw milk. Epidemiol. Infect. 1987; **98:** 263-269.

81. Hutchinson D N, Bolton F J, Hinchliff P M et al. Evidence of udder excretion of *Campylobacter jejuni* as the cause of milk-borne campylobacter outbreak. J. Hyg. 1985; **94:** 205-15.

82. Skirrow M B, Fidoe R G, Jones D M. An outbreak of presumptive foodborne *Campylobacter* enteritis. J. of Infect. 1981; **3:** 234-6.

83. Barrett N J. Communicable disease associated with milk and dairy products in England and Wales: 1983-84. J. Infect. 1986; **12:** 265-72.

84. Stehr-Green J, Mitchell P, Nicholls C, McEwan S, Payne A. *Campylobacter* enteritis - New Zealand. MMWR Morb. Mortal Wkly. 1991; **40:** 116-7.

85. Palmer S R, Gully P R, White J M. A waterborne outbreak of *Campylobacter* gastroenteritis. Lancet. 1983; **1:** 287-290.

86. Varnam A H, Evans M G. Foodborne pathogens: an illustrated text. London Wolfe Publishing Ltd. 1991; **222.**

87. Blaser M J, Checko P, Bopp C, Bruce A, Hughes J M. *Campylobacter* associated with foodborne transmission. Am. J. Epidemiol. 1982; **116:** 886-94.

88. Harris N V, Weiss N S, Nolan C M. The role of poultry and meats in the etiology of *Campylobacter jejuni/coli* enteritis. Am. J. Public Health 1986 Apr; **76(4):** 407-411.

89. Harris N V, Thompson D, Martin D C, Nolan C M. A survey of *Campylobacter* and other bacterial contaminants of pre-market chicken and retail poultry and meats, King County, Washington. Am. J. Public Health 1986 Apr; **76(4):** 401-406.

90. Hopkins R S, Scott A S. Handling raw chicken as a source for sporadic *Campylobacter jejuni* infections. J. Infect. Dis. 1983; **148:** 770.

91. Schmid G P, Schaefer R E, Plikaytis B D et al. A one-year study of endemic campylobacteriosis in a mid-western city: association with consumption of raw milk. J. Infect. Dis. 1987; **156:** 218-22

92. Humphrey T J, Hart R J C. *Campylobacter jejuni* in dairy cows and raw milk. J. Appl. Epidemiol. Infect. 1988; **65:** 463-467.

93. Aho M, Kurki M, Rautelin H, et al. Waterborne outbreak of *Campylobacter* enteritis after outdoor infantry drill in Utti, Finland. Epidemiol. Infect. 1989; **103:** 133-141.

94. Salfield N J, Pugh E J. *Campylobacter* enteritis in young children living in households with puppies. Br. Med. J. (Clin. Res. Ed.) 1987; 294 (6563), 21-2.

95. Skirrow M B. *Campylobacter* enteritis in dogs and cats: a 'new' zoonosis. Vet. Res. Commun. 1981; **5:** 13-19.

96. Seattle-King County Department of Public Health. Surveillance of the flow of *Salmonella* and *Campylobacter* in a community. Food and Drug Administration, Bureau of Veterinary Medicine, Washington, D C, 1984.

97. Park C E, Stankiewicz Z K, Lone H J. Incidence of *Campylobacter jejuni* in fresh and eviscerated whole market chickens. Can. J. Microbiol. 1981; **27:** 841-842.

98. Berndtson E, Tiremo M, Engrall A. Distribution and numbers of *Campylobacter* in newly slaughtered broiler chickens and hens. Int. J. Food Microbiol. 1992; **15:** 45-50.

99. Hood A M, Pearson A D, Shahamat M. The extent of surface contamination of retailed chickens with *Campylobacter jejuni* serogroups. Epidemiol. Infect. 1988; **100:** 17-25.

100. Stern N J, Green S S, Thaker N et al. Recovery of *Campylobacter jejuni* from fresh and frozen meat and poultry meat at slaughter. J. Food Prot. 1984; **47:** 372-374.

101. Pearson A D, Healing T D, Sockett P N. Surveillance of *Campylobacter* in England and Wales 1977-88: Is there an association between increasing human infection and consumption of fresh chicken? In: *Campylobacter* V. Editors: G M Ruiz-Palacios, E Calva, B R Ruiz-Palacios, Vasco de Quiroga, Mexico, 1991.

102. Appendix 2: *Campylobacter* enteritis. In: The Microbiological Safety of Food: Part I. Report of the Committee on the Microbiological Safety of Food (Chairman: Sir Mark Richmond). London: HMSO, 1990; 126-132.

103. Endtz H P, Ruijs G J, van Klingeren B, Jansen W H, van der Reyden T, Manton R P. Quinolone resistance in *Campylobacter* isolated from man and poultry following the introduction of fluroquinolones in veterinary medicine. J. Antimicrob. Chem 1991; **27:** 199-208.

104. de Boer E, Hahné M. Cross-contamination with *Campylobacter jejuni* and *Salmonella* species from raw chicken products during food preparation. J. Food Prot. 1990; **53(12):** 1067-1068.

105. Coates D, Hutchinson D N, Bolton F J. Survival of thermophilic campylobacters on fingertips and their elimination by washing and disinfection. Epidemiol. Infect. 1987; **99:** 265-274.

106. Blaser M J, La Force F M, Wilson N A, Wang W L L. Reservoirs for human campylobacteriosis. J. Infect. Dis. 1980; **141:** 665-9.

107. Reilly W J, Paterson G M, Sharp J C M. Milkborne Infection in Scotland. In: Proceedings of the Third World Congress Foodborne infections and Intoxications 85-88: Berlin 16-19 June. Robert von Ostertag-Institute, 1992.

108. Hudson S H, Sobo A O, Russell K et al. Jackdaws as a potential source of milkborne *Campylobacter jejuni* infection. Lancet. 1990; **335:** 1160.

109. Southern J P, Smith R M, Palmer S R. Bird attack on milk bottles; possible mode of transmission of *Campylobacter jejuni* to man. Lancet 1990; **336:** 1425-1427.

110. Lighton L L, Kaczmarski E B, Jones D M. A study of risk factors for *Campylobacter* infection in late spring. Public Health. 1991; **105:** 199-203.

111. Protecting milk quality on the doorstep. Milk Ind. 1991; **93(3):** 13.

112. Hudson S J, Lightfoot N F, Coulson J C, Russell K, Sisson P R, Obo AO. Jackdaws and Magpies as vectors of milkborne *Campylobacter* infection. Epidemiol. Infect. 1991; **107:** 363-372.

113. Pearson A D, Hooper W L, Lior H et al. Why investigate sporadic cases? The significance of fresh and New York dressed chicken as a source of campylobacter infection. In *Campylobacter* III. Editors: A D Pearson, M B Skirrow, H Lior, B Rowe. Public Health Laboratory Service. 1985; 290-291.

114. Taylor D N, McDermott K T, Little J R et al. *Campylobacter enteritis* associated with drinking untreated water in back country areas of the rocky mountains. Ann. Intern. Med. 1983; **99:** 38-40.

115. Jones K, Betaieb M, Telford D R. Thermophilic campylobacters in surface waters around Lancaster, UK: negative correlation with *Campylobacter* infections in the community. J. Appl. Bacteriol. 1990 Nov; **69(5):** 758-764.

116. Turnbull P C B, Rose P. *Campylobacter jejuni* and Salmonella in raw red meats. A Public Health Laboratory Service Survey. J. Hyg. (Camb). 1982; **88:** 29-37.

117. Banffer J R J. Biotype and serotype of *Campylobacter jejuni/coli* isolated from humans, pigs and chickens in the region of Rotterdam. Antonie van Leeuwenhoek 1985; **51(5/6):** 504-505.

118. Miller I S, Bolton F J, Dawkins H C. An outbreak of *Campylobacter* enteritis transmitted by puppies. Environ. Health. 1987; **95:** 11-14.

119. Blaser M J, Walderman R J, Barrett T, Erlandson A L. Outbreaks of *Campylobacter* enteritis in two extended families: evidence for person to person transmission. J. Paediatr. 1981; **98:** 254-257.

120. Cowden J M, Adak G K. National case-control study of primary indigenous sporadic cases of *Campylobacter* infection. Microb. Ecol. Health Dis. 1991; **4:** 571.

121. Smith T. Spirella associated with disease of the fetal membranes in cattle (infectious abortion). J. Exp. Med. 1918; **28:** 701-728.

122. Smith T, Taylor M S. Some morphological and biological characteristics of the *Spirilla* (*Vibrio fetus, n. sp.*) associated with disease of the foetal membranes in cattle. J. Exp. Med. 1919; **30:** 299-312.

123. Engvall A, Bergguist et al. Colonisation of broilers with *Campylobacter* in conventional broiler chicken flocks. Acta Veterinaria Scandiavica. 1986; **27:** 540 547.

124. Annan-Prah A, Janc M. The mode of spread of *Campylobacter jejuni/coli* to broiler flocks. J. Vet. Med. 1988; **35:** 11-18.

125. Shanker S, Lee A, Sorrell T C. Horizontal transmission of *Campylobacter jejuni* amongst broiler chicks: experimental studies. Epidemiol. Infect. 1990; **104:** 101-110.

126. Kaino K, Hayashidani H, Kanek O K, Ogawa M. Intestinal colonisation of *Campylobacter jejuni* in chickens. Jpn. J. Vet. Sci, 1988.

127. Shane S M, Gifford D H, Yogasundram K. *Campylobacter jejuni* contamination of eggs. Vet. Res. Commun. 1986; **10:** 487-492.

128. Neill S D, Campbell J N, Greene J A. *Campylobacter* species in broiler chickens. Avian Pathol. 1984; **13:** 777-785.

129. Van de Giessen A, Mazurier S I, Jacobs-Reitsma W, Janson W, Berkens P, Ritmeester W, Werneds K. Study on the epidemiology and control of *Campylobacter jejuni* in poultry broiler flocks. Appl. Environ. Microbiol. 1992; **58:** 1913-1917.

130. Hoop R, Ehrsam H. A contribution to the epidemiology of *Campylobacter jejuni/coli* in poultry meat production. Schweiz Arch Tierheilk 1987; **129:** 193-203.

131. Shane S M. The significance of *Campylobacter jejuni* infection in poultry: a review. Avian Path. 1992; **21:** 189-213.

132. Humphrey T J, Henley A, Lanning O G. The colonisation of broiler chickens with *Campylobacter jejuni*: some epidemiological investigations. Epidemiol. Infect. In Press, 1993.

133. Genigeorgis C, Hassuneh H & Collins P. *Campylobacter jejuni* infection on poultry farms and its effect on poultry meat contamination during slaughtering. J. Food Prot. 1986; **49:** 895-903.

134. Svedhem A, Kaijser B, Sjogren E. The occurrence of *Campylobacter jejuni* in fresh food and survival under different conditions. J. Hyg. (Lond) 1981 Dec; **87(3):** 421-425.

135. Kalenic S, Petrak T, Roseg D, Vodopija I. *Campylobacters* contaminate poultry carcases in industrial processing. In: *Campylobacter* IV, 298-299. Editors: B Kaijser, E Falson. Goteborg, Sweden, 1988.

136. Waterman S C, Park R W A, Bramley A J. A search for the source of *Campylobacter jejuni* in milk. J. Hyg. (Camb). 1984; **92:** 333-337.

137. Varga J, Mezes B, Fodor L, Hajtos I. Serogroups of *Campylobacter fetus* and *Campylobacter jejuni* isolated in cases of ovine abortion. J. Vet. 1990; Med. **37:** 148-152.

138. Terzolo H R. Identification of campylobacters from bovine and ovine faeces. Rev. Argent. Microbiol. 1988 Apr-Jun; **20(2):** 53-68.

139. Gebhart C J, Lin G F, McOrist S M, Lawson G H K & Murtaugh M P. Cloned DNA probes specific for the intracellular campylobacter-like organisms of porcine proliferative enteritis. J. Clin. Microbiol. 1991; **29:** 1011-1015.

140. Bolton F J, Dawkins H C, Hutchinson D N. Biotypes and serotypes of thermophilic campylobacters isolated from cattle, sheep and pig offal and other red meats. J. Hyg. (Lond) 1985; **95:** 1-6.

141. Bruce D, Zochowski W, Fleming G A. *Campylobacter* species in cats and dogs. Vet. Rec. 1980; **107:** 200-201.

142. Prescott J F, Monroe D L. *Campylobacter jejuni* enteritis in man and domestic animals. J. Am. Vet. Med. Assoc. 1982; **181:** 1524-1530.

143. Blankenship L C, Craven S E. *Campylobacter jejuni* survival in chicken meat as a function of temperature. Appl. Environ. Microbiol. 1982; Jul; **44(1):** 88-92.

144. Gill G O, Harris L M. Survival and growth of *Campylobacter fetus* subsp. *jejuni* on meat and cooked foods. Appl. Environ. Microbiol. 1982; **44:** 259-263.

145. Hänninen M L. Growth and survival characteristics of *Campylobacter jejuni* in liquid egg. J. Hyg. Camb. 1984; **92:** 53-58.

146. Hänninen M L. Effect of various gas atmospheres on the growth and survival of *Campylobacter jejuni* on beef. J. Appl. Bact. 1984; **57:** 89-94.

147. Clarke A G, Bueschkens D H. Survival and growth of *Campylobacter jejuni* in egg yolk and albumen. J. Food Prot. 1986; **49:** 135-141.

148. Christopher F M, Smith G C, Vanderzant C. Effect of temperature and pH on the survival of *Campylobacter* fetus. J. Food Prot. 1982; **45(3):** 253-259.

149. Blaser M J, Perez G P, Smith P F, Patton C, Tenover F C, Lastovica A J. Extraintestinal *Campylobacter jejuni* and *Campylobacter coli* infections: host factors and strain characteristics. J. Infect. Dis. 1986; **153:** 552-559.

150. Rosef O, Kapperud G. Isolation of *Campylobacter fetus* subsp. *jejuni* from faeces of Norwegian poultry. Acta Vet. Scand. 1982; **23(1):** 128-34.

151. Gill K P, Bates P G, Lander K P. The effect of pasteurization on the survival of *Campylobacter* species in milk. Brit. Vet. J. 1981 Nov; **137(6):** 378-384.

152. Stern N J, Kotula A W. Survival of *Campylobacter jejuni* inoculated into ground beef. Appl. Environ. Microbiol. 1982. Nov; **44(5):** 1150-1153.

153. Doyle M P, Roman D J. Response of *Campylobacter jejuni* to sodium chloride. Appl. Environ. Microbiol. 1982 Mar; **43(3):** 561-565.

154. Shapton D A. Biological factors underlying food safety, preservation and stability. In: Shapton DA, Shapton NF. Principles and practices for the safe processing of foods. Oxford: Butterworth-Heinemann, 1991; 222-253.

155. Hudson W R, Roberts T A. The occurrence of *Campylobacter jejuni* on commercial red meat carcases from an abbatoir. In: Newell DC, editors. *Campylobacter*: epidemiology, pathogenesis and biochemistry. Lancaster: MTP Press Ltd, 1982; 273.

156. Doyle M P, Roman D J. Growth and survival of *Campylobacter fetus* subsp. *jejuni* as a function of temperature and pH. J. Food Prot. 1981; **44(8):** 596-601.

157. Farber J M. Microbiological aspects of modified atmosphere packaging technology -A review. J. Food Prot. 1991; **54(1):** 58-70.

158. Phebus R K, Dranghon F A, Mount J R. Survival of *Campylobacter jejuni* in modified atmosphere packaged turkey roll. J. Food Prot. 1991; **54:** 194-199.

159. Lambert J D, Maxey R B. Effect of gamma radiation on *Campylobacter jejuni*. J. Food Sci. 1984, **49:** 665-667, 674.

160. Patterson M. Sensitivity of bacteria to irradiation in poultry meat under various atmospheres. Letters Applied Microbiol. 1988; **7:** 55-58.

161. Mitchell B. How to HACCP. Br. Food J. 1992; **94(1):** 16-20.

162. Campden Food and Drink Research Association. HACCP: A practical guide. Chipping Campden: Campden Food and Drink Research Association, 1992.

163. Ministry of Agriculture Fisheries and Food. Food Safety [Leaflet]. London: Foodsense, 1991.

164. Food Hygiene (Amendment) Regulations. London: HMSO, 1990.

165. Food Hygiene (Amendment) Regulations. London: HMSO, 1991.

166. Evans J A, Stanton J I, Russell S L, James S J. Consumer handling of chilled foods: a survey of time and temperature conditions London: Ministry of Agriculture, Fisheries and Food; 1991.

167. Safer cooked meat production guidelines. DH/London: Department of Health, 1992.

168. Dairy Trade Federation. Guidelines for the good hygiene practice in the manufacture of dairy-based products. London: Dairy Trade Federation, 1989.

169. Lelieveld H L M, Hugelshofer W, Jepson P C, Lalande M, Mostert M A, Nassauer J. Safe pasteurisation. Lebensmitteltechnik 1992; **24(7/8):** 38-43.

170. Department of Health. Chilled and frozen; guidelines on the cook-chill and cook-freeze catering systems. London: HMSO, 1989.

171. Karmali M A, Allen A K, Fleming P C. Differentiation of catalase positive *Campylobacters* with special reference to morphology. Int. J. Syst. Bacteriol. 1981; **31:** 64-71.

172. Moran A P, Upton M E. Factors affecting production of coccoidal forms by *Campylobacter jejuni* in solid media during incubation. J. Appl. Bacteriol. 1987; **62:** 527-537.

173. Steele T W, Owen R J. *Campylobacter jejuni subsp.doylei subsp. nov.*, a subspecies of nitrate negative *Campylobacters* isolated from human clinical specimens. Int. J. Syst. Bacteriol. 1988; **38:** 316-318.

174. Jones, D M, Sutcliffe, E M, Rias R, Fox A J, Curry A. *Campylobacter jejuni* adapts to aerobic metabolism in the environment. J. of Med. Microbiol. 1993; **38:** 145-150.

175. Burnens A P, Nicolet J. Three supplementary diagnostic tests for *Campylobacter* species and related organisms. J Clin. Microbiol; 1993; **31:** 708-710.

176. Stanley J, Burnens A P, Linton D, On S L, Costas M, Owen R J. *Campylobacter helvetious* sp. nov. a new thermophilic species from domestic animals; characterisation and closing of a new species - specific DNS probe. J. Gen. Microbiol. 1992; **138:** 2293-2303.

177. Goosens H, Vlaes L, De Boeck M, Pot B, Kersters K, Levy J, De Moe P, Butzler J P. Is *Campylobacter upsaliensis* an unrecognised cause of human diarrhoea? Lancet 1990; **335:** 584.

178. Lastovica A J, Le Roux E. Prevalence and distribution of *Campylobacter* species in paediatric patients. Microb. Ecol. Health Dis. 1991; **4:** 586.

179. Lauwers S T, Devrekor T, Van Etteryck P, Breynaert J, Van Zeebroeck A, Smekens L et al. Isolation of *Campylobacter* concisus from human faeces. Microb. Ecol. Health Dis. 1991; **4:** 591.

180. Pugina, P, Benzi, G, Lauwers, S, Van Elterijck, R, Butzler, J P, Vlaes, L, Vandamme, P. An outbreak of "Arcobacter (*Campylobacter*) butzleri" in Italy. In: The IVth International Workshop on *Campylobacter, Helicobacter* and related organisms. Microbial. Ecology in Health and Disease, 4, special issue, 1991.

181. Vandamme P, Vanconneyt M, Pot B, Mels L, Haste B, Dewettinck D, Vlaes L, Canden Borre C, Higgins R. Genus arcobacter with *Arcobacter butzteri* comb. Nov. and Arcobacter skirrowii sp nov. an aerotolerant bacterium isolated from veterinary specimens. Int. J. Syst. Bacteriol. 1992; **42:** 344-356.

182. Vandamme P, Delay J. Proposal for a new family, Campylobacteraceae. Int. J. Syst. Bacteriol. 1993; **41:** 451.

183. Jones D M, Abbot J D, Painter M J, Sutcliffe E M. A comparison of biotypes and serotypes of *Campylobacter* sp. isolated from patients with enteritis and from animal and environmental sources. J. Infect. 1984; **9:** 51-58.

184. King E O. Human infections with *Vibrio fetus* and a closely related vibrio. J. Infect. Dis. 1957; **101:** 119-128.

185. Nachamkin I. *Campylobacter* infections. Current Sciences 1993; 72-76.

186. Endtz H P, Ruijs G J, Zwinderman A H, van der Reyden T, Bierer M, Manton R P. Comparison of six media, including a semi-solid agar, for the isolation of various *Campylobacter* species from stool specimens. J. Clin. Microbiol. 1991; **29:** 1007-1010.

187. Rogol M, Shpak B, Rothman D, Sechter I. Enrichment medium for the isolation of *Campylobacter jejuni* and *Campylobacter coli*. Appl. Environ. Microbiol. 1985; **50:** 125-126.

188. Hutchinson D N, Bolton F J. Is enrichment culture necessary for the isolation of *Campylobacter jejuni* from faeces? J. Clin. Pathol. 1983 Dec; **36(12):** 1350-1352.

189. Bolton F J, Robertson L. A selective medium for isolating *Campylobacter jejuni/coli*. J. Clin. Pathol. 1982; Apr; **35(4):** 462-467.

190. Hutchinson D N, Bolton F J. Improved blood-free selective medium for the isolation of *Campylobacter jejuni* from faecal specimens. J. Clin. Pathol. 1984; **37:** 956-957.

191. Goosens H, de Boeck M, Butzler J P. A new selective medium for the isolation of *Campylobacter jejuni* from human faeces. European J. Clin. Microbiol. 1983; **2:** 389-394.

192. Griffiths P L, Park R W A. Campylobacters associated with human diarrhoeal disease. J. Appl. Bacterial. 1990; **69:** 281-301.

193. Goosens H, Vlaes L, Galand I, Van den Borre C, Butzler J P. A new semi-solid selective medium for the isolation of *Campylobacter* stool specimens. In: *Campylobacter* IV, editors G M Ruiz-Palacios, E Calua, B R Ruiz-Palacios, Vasco de Quirosa, Mexico, Department of Infectious Diseases. 1991; 103-106.

194. Kiehlbauch J A, Baker C N, Wachsmuth I K. In vitro susceptibilities of aerotolerant *Campylobacter* isolates to 22 anti-microbiol. agents. Antimicrob. Agents Chemother. 1992; **36:** 717.

195. Saha S K, Saha S, Sanyal S C. Recovery of injured *Campylobacter jejuni* cells after animal passage. Appl. Environ. Microbiol. 1991; **57:** 3388-3389.

196. Beumer R R, de Vries J, Rombouts F M. *Campylobacter jejuni* non-culturable coccoid cells. Int. J. Food Microbiol. 1992; **15:** 153-63.

197. Taylor D E, Hiratsuka K. Use of non-radioactive DNA probes for detection of *Campylobacter jejuni* and *C.coli* in stool specimens. Mol. Cell Probes 1990; **4:** 261-271.

198. Tenover F C, Carlson L, Barbagallo S, Nachamkin I. DNA probe culture confirmation assay for identification of thermophilic *Campylobacter* species. J. Clin. Microbiol. 1990; **28:** 1284-1287.

199. Oyofo B A, Thornton S A, Burr D H, Trust T J, Pavlovskiso, Guerry P. Detection of *Campylobacter jejuni* and *Campylobacter coli* using the polymerase chain reaction. In: abstracts from the 92nd general meeting of the American Society for Microbiology 26-30 May 1992. New Orleans: American Society for Microbiology. 1992; **34**.

200. Roche E S, Weiss J B. Detection and differentiation of *Campylobacter* species using polymerase chain reaction. In: 31st international conference on antimicrobial agents and chemotherapy. Chicago USA, 1991.

201. Giesendorf B A J, Quint W G V, Henkens M H C, Stegeman H, Huf F A, Niesters H G M. Rapid and sensitive detection of *Campylobacter* species in chicken products by using the Polymcrase chain reaction. Appl. Environ. Microbiol. 1992; **58:** 3804-3808.

202. Skirrow M B, Benjamin J. '1001' *Campylobacters*: cultural characteristics of intestinal *Campylobacters* from man and animals. J. Hyg. (Camb) 1980; 427-442.

203. Skirrow M B, Benjamin J. Differentiation of enteropathogenic *Campylobacters*. J. Clin. Pathol. 1980; **33:** 1122.

204. Lior H. New, extended biotyping scheme for *Campylobacter jejuni*, *Campylobacter coli*, and *Campylobacter laridis*. J. Clin. Microbiol. 1984; **20:** 636-640.

205. Roop R M, Smibert R M, Krieg N R. Improved biotyping schemes for *Campylobacter jejuni* and *Campylobacter coli*. J. Clin. Microbiol. 1984; **20:** 990-992.

206. Khakhria R, Lior H. Extended phage-typing scheme for *Campylobacter jejuni* and *Campylobacter coli*. Epidemiol. Infect. 1992; **108:** 403-14.

207. Jones, D M, Sutcliffe, E M. Serotypes of thermophilic campylobacters isolated in Manchester, UK over seven years. In: The Vth International Workshop on *Campylobacter* infections. Ed: M Ruiz-Palacios, Puerto Vallarta, Mexico, 1989.

208. Nicholson M A, Patton C M. Evaluation of commercial antisera for serotyping heat-labile antigens of *Campylobacter jejuni* and *Campylobacter coli*. J. Clin. Microbiol. 1993; **31:** 900-903.

209. Lior, H, Woodward, D L. Serotyping and biotyping of *Campylobacter*, an update. In: *Campylobacter* IV, Editors: B. Kaijser and E. Falsen. Dept. Clinical Bacteriology, University of Göteborg, Sweden, 1987.

210. Hebert G A, Hollis D G, Weaver R E, Steigerwalt A G, McKinney R M, Brenner D J. Serogroups of *Campylobacter/coli*, and *Campylobacter fetus* defined by direct immunofluorescence. J. Clin. Microbiol. 1983; **17:** 529-538.

211. Abbot J D, Dale B, Eldridge J, Jones D M. Serotyping of *Campylobacter jejuni/coli*. J. Clin. Pathol. 1980; **33:** 762-766.

212. Patton C M, Barrett T J, Morris G K. Comparison of the Penner and Lior methods for serotyping *Campylobacter* species. J. Clin. Microbiol. 1985; **22:** 558-565.

213. Morris J A, Park R W A. A comparison using gel electrophoresis of cell proteins of *Campylobacters* (Vibrios) associated with infertility, abortion and swine dysentery. J. Gen. Microbiol. 1973; **78:** 165-178.

214. Ferguson D A, Lambe D W. Differentiation of *Campylobacter* species by protein banding patterns in polyacrylamide slab gels. J. Clin. Microbiol. 1984; **20:** 453-460.

215. Owen R J, Costas M, Sloss L, Lastovica A, Le-Roux E. Identification of catalase negative/weak *Campylobacter jejuni* from human blood and faecal cultures by numerical analysis of electropheretic protein patterns. FEMS-Microbiol-Lett 1990; **57:** 329-335.

216. Owen R J, Costas M, Dawson C. Application of different chromosomal DNA fingerprints to specific and subspecific identification of *Campylobacter* isolates. J. Clin. Microbiol. 1989; **27:** 2338-2343.

217. Owen R J, Hernandez J, Bolton F. DNA restriction digest and ribosomal RNA gene patterns of *Campylobacter jejuni*: a comparison with bio-, ser-, and bacteriophage-types of United Kingdom outbreak strains. Epidemiol. Infect. 1990; **105:** 265-275.

218. Fayos A, Owen R J, Desai M, Hernandez J. Ribosomal RNA gene restriction fragment diversity amongst Lior biotypes and Penner serotypes of *Campylobacter jejuni* and *Campylobacter coli*. FEMS Microbiol. Lett. 1992; **74:** 87-93.

219. Moureau P, Derclaye I, Gregoire D, Janssen M, Cornelis G R. *Campylobacter* species identification based on polymorphism of DNA encoding RNA. J. Clin. Microbiol. 1989; **27:** 1514-1517.

220. Taylor D E, Eaton M, Yan W, Chang N. Genome maps of *Campylobacter jejuni* and *Campylobacter coli*. J. Bacteriol. 1992; **174:** 2332-2337.

221. Yan W, Chang N, Taylor D E. Pulsed field gel electrophoresis of *Campylobacter jejuni* and *Campylobacter coli* genomic DNA and its epidemiologic application. J. Int. Dis. 1991; **163:** 1068-1072.

222. Aeschbacher M, Piffaretti J-C. Population genetics of human and animal enteric *Campylobacter* strains. Infect. Immun. 1989; **57:** 1432-1437.

223. Saiki, R K, Scharf, S J, Faloona, F, Mullis, K B, Horn, G T, Erlich, H A, Arnheim, N. Enzmatic amplification of Beta globulin genomic sequences and restriction site analysis for diagnosis of sickle cell anaemia. Science. 1985; **230:** 1350-1354.

224. Mazurier S, van de Giesson A, Heuvelman K, Wesnier K. RAPD analysis of *Campylobacter* isolates: DNA fingerprinting without the need to purify DNA. Lett. Appl. Microbiol. 1992; **14:** 260-262.

225. Nuijten P J M, Bleumink-Pluym N M C, Gaastra W, Van der Zeijst B A M. Flagellin expression in *Campylobacter jejuni* is regulated at the transcriptional level. Infect. Immun. 1989; **57:** 1084-1088.

226. Lee A, O'Rouke J L, Barrington P J, Trust T J. Mucus colonisation as a determinant of pathogenicity in intestinal infection by *Campylobacter jejuni*: a mouse cecal model. Infect. Immun. 1986; **51:** 536-546.

227. Newell D G, and Pearson A D. Pathogenicity of *Campylobacter jejuni*- an *in vitro* model of adhesion and invasion? In: *Campylobacter*; epidemiology, pathogenesis and biochemistry. Editor; D G Newell. MTP Press. Lancaster, Boston, The Hague, 1981.

228. McBride H, Newell D G. *In vitro* models of adhesion for *Campylobacter jejuni*. In; *Campylobacter* II. Editor; A D Pearson, M B Skirrow, B Rowe, J R Davies, and D M Jones. Public Health Laboratory Service. London, 1983.

229. McSweegan E, Walker R. Isolation and characterisation of two *Campylobacter jejuni* adhesins for cellular and mucous substrates. Infect. Immun. 1986; **53:** 141-148.

230. Konkel M E, Joens L A. Adhesion to and invasion of HEp cells by *Campylobacter* species. Infect. Immun. 1989; **57:** 2984-2990.

231. De Melo M A, Gabbiani G, Pechere J-C. Cellular events and intracellular survival of *Campylobacter jejuni* during infection of Hep 2 cells. Infect. Immun. 1989; **57:** 2214-2222.

232. Konkel M E, Hayes S F, Joens L A, Gieplak W J. Characteristics of the internalisation and intracellular survival of *Campylobacter jejuni* in human epithelial cell cultures. Microb. Pathog. 1992; **13:** 357-370.

233. Newell D G, McBride H, Sauders F, Dehele Y, Pearson A D. Virulance of clinical and environmental isolates of *Campylobacter jejuni*. J. Hyg. (Camb.) 1985; **94:** 45-54.

234. King V, Wassenaar T, Van der Zeijst B A M, Newell D G. Variations in *Campylobacter jejuni* flagellin and flagellin genes during in vivo and in vitro passage. Microbiol Ecol. Health Dis. 1991; **101:** 119-128.

235. Konkel M E, Babakhani F, Joens L A. Invasion related antigens of *Campylobacter jejuni*. J. Infect. Dis. 1990; **162:** 888-895.

236. McCardell B A, Madden J M, Lee E C. Production of cholera-like toxin by *Campylobacter jejuni/coli* (letter). Lancet 1984 Feb 25; **1(8374):** 448-449.

237. Klipstein E A, Engert R E, Short H, Schenk E A. Pathogenic properties of *Campylobacter jejuni*: assay and correlation with clinical manifestations. Infect. Immun. 1985; **50:** 43-49.

238. Johnson W M, Lior H. A new heat-labile cytolethal distending toxin (CLDT) produced by *Campylobacter* species. Microbiol. Path. 1988; **4:** 115-126.

239. Florin I, Antillon F. Production of enterotoxin and cytotoxin in *Campylobacter* strains isolated in Costa Rica. J. Med. Microbial. 1992; **37:** 22-9.

240. Perez-Perez G I, Cohn D L, Guerrant R L, Patton C M, Reller L B, Blaser M J. Clinical and immunologic significance of cholera-like toxin and cytotoxin production by *Campylobacter* species in patients with inflammatory diarrhoea in the USA. J. Infect. Dis. 1989; **160:** 460-468.

241. Lee E, McCardell B, Guerry P. Characterisation of a plasmid-encoded enterotoxin in *Campylobacter jejuni*. In: *Campylobacter* III, Editors: A D Pearson, M B Skirrow, H Lior, B Rowe. PHLS, London, 1985.

242. Watson K C, Kerr E J, McFazdean S M. Serology of *Campylobacter* infections. J. Infect. 1979; **1:** 151-158.

243. Jones D M, Eldridge J, Dale B. Serological response of *Campylobacter jejuni/coli* infection. J. Clin. Pathol. 1980; **33:** 767-769.

244. Svedhem A, Gunnerson H, Kaijser B. In; *Campylobacter*; epidemiology, pathogenesis, and biochemistry. Editor; D G Newell. MTP Press, Lancaster, Boston, The Hague; 1982.

245. Nachamkin I, Hart A M. Western blot analysis of the human antibody reponse to *Campylobacter jejuni* cellular antigens during gastrointestinal infection. J. Clin. Microbiol. 1985; **21:** 33-38.

246. Nachamkin I, Yang X H. Human antibody response to *Campylobacter jejuni* flagellin protein and a synthetic N-terminal peptide. J. Clin. Microbiol. 1989; **27:** 2195-2198.

247. Perlman D M, Ampel N M, Schifman R B, Cohn D L, Patton C M, Aguirre ML et al. Persistent *Campylobacter jejuni* infections in patients infected with the human immunodeficiency virus (HIV). Ann. Intern. Med. 1988 Apr; **108(4):** 540-546.

248. Christenson B, Ringner A, Blucher C, Billaudelle H, Gundtoft K N, Eriksson G, Bottiger M. An outbreak of *Campylobacter* enteritis among the staff of a poultry abattoir in Sweden. Scand. J. Infect. Dis. 1983; **15:** 167-172.

249. Ruiz-Palacios G M, Cervantes L E, Soto L E, Newberg P W, Pickering L K. *Campylobacter jejuni* receptor analogues are present in human milk. Pediatr. Res. 1992; **31** (4 pt 2), 178A.

250. McSweegan E, Burr D H, Walker R I. Intestinal mucus gel and secretory antibody are barriers to *Campylobacter* adherence to INT 407 cells. Infect. Immun. 1987; **55:** 1431-1435.

251. Abimiku AG, Dolby JM. Cross-protection of infant mice against intestinal colonisation by *Campylobacter jejuni*: importance of heat-labile serotyping (Lior) antigens. J. Med. Microbiol. 1988; Aug; **26(4):** 265-268.

Printed in the United Kingdom for HMSO.
Dd.0297068, C15, 9/93, 3396/4, 5673, 259370.